Chaotic Maps

Dynamics, Fractals, and Rapid Fluctuations

Synthesis Lectures on Mathematics and Statistics

Editor
Steven G. Krantz, *Washington University, St. Louis*

Jordan Canonical Form: Application to Differential Equations
Steven H. Weintraub
2008

Statistics is Easy!
Dennis Shasha and Manda Wilson
2008

A Gyrovector Space Approach to Hyperbolic Geometry
Abraham Albert Ungar
2008

Chaotic Maps: Dynamics, Fractals, and Rapid Fluctuations

Goong Chen and Yu Huang

ISBN: 978-3-031-01275-4   paperback
ISBN: 978-3-031-02403-0   ebook

DOI 10.1007/978-3-031-02403-0

A Publication in the Springer series
SYNTHESIS LECTURES ON MATHEMATICS AND STATISTICS

Lecture #11
Series Editor: Steven G. Krantz, *Washington University, St. Louis*
Series ISSN
Synthesis Lectures on Mathematics and Statistics
Print 1938-1743   Electronic 1938-1751

Chaotic Maps

Dynamics, Fractals, and Rapid Fluctuations

Goong Chen
Texas A&M University

Yu Huang
Sun Yat-Sen University

SYNTHESIS LECTURES ON MATHEMATICS AND STATISTICS #11

ABSTRACT

This book consists of lecture notes for a semester-long introductory graduate course on dynamical systems and chaos taught by the authors at Texas A&M University and Zhongshan University, China. There are ten chapters in the main body of the book, covering an elementary theory of chaotic maps in finite-dimensional spaces. The topics include one-dimensional dynamical systems (interval maps), bifurcations, general topological, symbolic dynamical systems, fractals and a class of infinite-dimensional dynamical systems which are induced by interval maps, plus rapid fluctuations of chaotic maps as a new viewpoint developed by the authors in recent years. Two appendices are also provided in order to ease the transitions for the readership from discrete-time dynamical systems to continuous-time dynamical systems, governed by ordinary and partial differential equations.

KEYWORDS

chaos, interval maps, periodicity, sensitive dependence, stability, Sharkovski's theorem, bifurcations, homoclinicity, symbolic dynamics, smale horseshoe, total variations, rapid fluctuations, fractals, wave equation

Contents

Preface

The understanding and analysis of chaotic systems are considered as one of the most important advances of the 20th Century. Such systems behave contrary to the ordinary belief that the universe is always orderly and predicable as a grand ensemble modelizable by differential equations. The great mathematician and astronomer Pierre-Simon Laplace (1749-1827) once said:

> "We may regard the present state of the universe as the effect of its past and the cause of its future. An intellect which at a certain moment would know all forces that set nature in motion, and all positions of all items of which nature is composed, if this intellect were also vast enough to submit these data to analysis, it would embrace in a single formula the movements of the greatest bodies of the universe and those of the tiniest atom; for such an intellect nothing would be uncertain and the future just like the past would be present before its eyes." (Laplace, *A Philosophical Essay on Probabilities* [47].)

Laplace had a conviction that, knowing all the governing differential equations and the initial conditions, we can predict everything in the universe in a deterministic way. But we now know that Laplace has underestimated the complexities of the equations of motion. The truth is that rather simple systems of ordinary differential equations can have behaviors that are extremely sensitive to initial conditions as well as manifesting randomness. In addition, in quantum mechanics, even though the governing equation, the Schrödinger equation, is deterministic, the outcomes from measurements are probabilistic.

The term "chaos," literally, means confusion, disarray, disorder, disorganization. turbulence, turmoil, etc. It appears to be the antithesis of beauty, elegance, harmony, order, organization, purity, and symmetry that most of us are all indoctrinated to believe that things should rightfully be. And, as such, chaos seems inherently to defy an organized description and a systematic study for a long time. Henri Poincaré is most often credited as the founder of modern dynamical systems and the discoverer of chaotic phenomena. In his study of the three-body problem during the 1880s, he found that celestial bodies can have orbits which are nonperiodic, and yet for any choices of period of motion, that period will not be steadily increasing nor approaching a fixed value. Poincaré's interests have stimulated the development of *ergodic theory*, studied and developed by prominent mathematicians G.D. Birkhoff, A.N. Kolmogorov, M.L. Cartwright, J.E. Littlewood, S. Smale, etc., mainly from the nonlinear differential equations point of view.

With the increasing availability of electronic computers during the 1960s, scientists and engineers could begin to play with them and in so doing have discovered phenomena never known before. Two major discoveries were made during the early 1960s: the *Lorenz Attractor*, by Edward Lorenz in his study of weather prediction, and *fractals*, by Benoît Mandelbrot in the study of fluctuating cotton prices. These discoveries have deeply revolutionized the thinking of engineers, mathematicians, and scientists in problem solving and the understanding of nature and shaped the future directions in the research and development of nonlinear science.

The actual coinage of "chaos" for the field is due to a 1975 paper by T.Y. Li and J.A. Yorke [49] entitled *Period Three Implies Chaos*. It is a perfect, captivating phrase for a study ready to take off, with enthusiastic participants from all walks of engineering and natural and social sciences. More rigorously speaking, a system is said to be *chaotic* if

(1) it has sensitive dependence on initial conditions;

(2) it must be topologically mixing; and

(3) the periodic orbits are dense.

Several other similar, but non-equivalent, definitions are possible and are used by different groups.

Today, *nonlinear science* is a highly active established discipline (and interdiscipline), where bifurcations, chaos, pattern formations, self-organizations, self-regulations, stability and instability, fractal structures, universality, synchronization, and peculiar nonlinear dynamical phenomena are some of the most intensively studied topics.

The topics of dynamical systems and chaos have now become a standard course in both the undergraduate and graduate mathematics curriculum of most major universities in the world. This book is developed from the lecture notes on dynamical systems and chaos the two authors taught at the Mathematics Departments of Texas A&M University and Zhongshan (Sun Yat-Sen) University in Guangzhou, China during 1995-2011.

The materials in the notes are intended for a semester-long introductory course. The main objective is to familiarize the students with the theory and techniques for (discrete-time) maps from mainly an analysis viewpoint, aiming eventually to also provide a stepping stone for nonlinear systems governed by ODEs and PDEs. The book is divided into ten chapters and two appendices. They cover the following major themes:

(I) **Interval maps**: Their basic properties (Chapter 1), Sharkovski's Theorem on periodicities (Chapter 3), bifurcations (Chapter 4), and homoclinicity (Chapter 5).

(II) **General dynamical systems and Smale Horseshoe**: The 2- and k-symbol dynamics, topological conjugacy and shift invariant sets (Chapter 6), and the Smale Horseshoe (Chapter 7).

(III) **Rapid fluctuations and fractals**: Total variations and heuristics (Chapter 2), fractals (Chapter 8), and rapid fluctuations of multi-dimensional maps and infinite-dimensional maps (Chapters 9 and 10).

Appendix A is provided in order to show some basic qualitative behaviors of higher-dimensional differential equation systems, and how to study continuous-time dynamical systems, which are often described by nonlinear ordinary differential equations, using its *Poincaré section*, which is a *map*. This would, hopefully, give the interested reader some head start toward the study of continuous-time dynamical systems.

Appendix B offers an example of a concrete case of an infinite-dimensional system described by the one-dimensional wave equation with a van der Pol type nonlinear boundary condition, and shows how to use interval maps and rapid fluctuations to understand and prove chaos.

For these three major themes (I)–(III) above, much of the contents in (I) and (II) are rather standard. But the majority of the materials in Theme (III) is taken from the research done by the two authors and our collaborators during the recent years. This viewpoint of regarding chaos as exponential growth of total variations on a strange attractor of some fractional Hausdorff dimensions is actually mostly stimulated by our research on the chaotic vibration of the wave equation introduced in Appendix B.

There are already a good number of books and monographs on dynamical systems and chaos on the market. In developing our own instructional materials, we have referenced and utilized them extensively and benefited immensely. We mention, in particular, the excellent books by Afraimovich and Hsu [2], Devaney [20], Guckenheimer and Holmes [30], Meyer and Hall [54], Robinson [58], and Wiggins [69]. In addition, we have also been blessed tremendously from two Chinese sources: Wen [67] and Zhou [75]. To these book authors, and in addition, our past collaborators and students who helped us either directly or indirectly in many ways, we express our sincerest thanks.

Professor Steven G. Krantz, editor of the book series, and Mr. Joel Claypool, book publisher, constantly pushed us, tolerated the repeated long delays, but kindly expedited the publication process. We are truly indebted.

The writing of this book was supported in part by the Texas Norman Hackman Advanced Research Program Grant #010366-0149-2009 from the Texas Higher Education Coordinating Board, Qatar National Research Fund (QNRF) National Priority Research Program Grants #NPRP09-462-1-074 and #NPRP4-1162-1-181, and the Chinese National Natural Science Foundation Grants #10771222 and 11071263.

Goong Chen and Yu Huang
July 2011

CHAPTER 1

Simple Interval Maps and Their Iterations

The discovery of many nonlinear phenomena and their study by systematic methods are a major breakthrough in science and mathematics of the 20th Century, leading to the research and development of *nonlinear science*, which is at the forefront of science and technology of the 21st Century. Chaos is an extreme form of nonlinear dynamical phenomena. But what exactly is *chaos*? This is the main focus of this book.

Mathematical definitions of chaos can be given in many different ways. Though we will give the first of such definitions at the end of Chapter 2 (in Def. 2.7), during much of the first few chapters the term "chaos" (or its adjective "chaotic") should be interpreted in a rather liberal and intuitive sense that it stands for some irregular behaviors or phenomena. This vagueness should automatically take care of itself once more rigorous definitions are given.

1.1   INTRODUCTION

We begin by considering some population models. A simple one is the Malthusian law of linear population growth:

$$\begin{cases} x_0 > 0 \quad \text{is given;} \\ x_{n+1} = \mu x_n; \qquad n = 0, 1, 2, \ldots \end{cases} \tag{1.1}$$

where

$$x_n = \text{the population size of certain biological species at time } n,$$

and $\mu > 0$ is a constant. For example, $\mu = 1.03$ if

$$0.03 = 3\% = \text{net birth rate} = \text{birth rate} - \text{death rate}.$$

The solution of (1.1) is

$$x_n = \mu^n x_0, \qquad n = 1, 2, \ldots .$$

Therefore,

$$\begin{cases} x_n \to \infty, & \text{if } \mu > 1, \text{ as } n \to \infty, \\ x_n = x_0, & \forall n = 1, 2, \ldots, \text{ if } \mu = 1, \\ x_n \to 0, & \text{if } \mu < 1, \text{ as } n \to \infty. \end{cases} \tag{1.2}$$

Thus, the long-term, or *asymptotic behavior*, of the system (1.1) is completely answered by (1.2). The model (1.1) from the population dynamics point of view is quite naive. An improved model of (1.1) is the following:

$$\begin{cases} x_{n+1} = \mu x_n - a x_n^2, \\ x_0 > 0 \quad \text{is given,} \end{cases} \qquad a > 0, \qquad n = 0, 1, 2, \ldots \qquad (1.3)$$

where the term $-a x_n^2$ models conflicts (such as competition for the same resources) between members of the species. It has a negative effect on population growth. Equation (1.3) is called the *modified Malthusian law* for population growth.

For a non-linear system with a single power law non-linearity, we can always "scale out" the coefficient associated with the non-linear term. Let $x_n = k y_n$ and substitute it into (1.3). We obtain:

$$k y_{n+1} = \mu(k y_n) - a(k y_n)^2$$
$$y_{n+1} = \mu y_n - a k y_n^2.$$

Set $k = \mu/a$. We have:

$$y_{n+1} = \mu y_n - \mu y_n^2 = \mu y_n(1 - y_n).$$

Rename $y_n = x_n$. We obtain

$$x_{n+1} = \mu x_n(1 - x_n) \equiv f(\mu, x_n), \qquad (1.4)$$

where

$$f(\mu, x) = f_\mu(x) = \mu x(1 - x). \qquad (1.5)$$

The map f is called the *quadratic map* or *logistic map*. It played a very important role in the development of chaos theory due to the study of the British biologist, Robert May, who noted (1975) that as μ changes, the system does not attain simple steady states as those in (1.2). One of our main interests here is to study the *asymptotic behavior* of the iterates of (1.4) as $n \to \infty$. Iterations of the type $x_{n+1} = f(x_n)$ happen very often elsewhere in applications, too. We look at another example below.

Example 1.1   Newton's algorithm for finding the zero of a given function $g(x)$.

Newton's algorithm provides a useful way for approximating solutions of an equation $g(x) = 0$ iteratively. Start from an initial point x_0, we compute $x_1, x_2, \ldots$, as follows. At each point x_n, draw a tangent line to the curve $y = g(x)$ passing through $(x_n, g(x_n))$:

$$y - g(x_n) = g'(x_n)(x - x_n).$$

This line intersects the x-axis at $x = x_{n+1}$:

$$0 - g(x_n) = g'(x_n)(x_{n+1} - x_n).$$

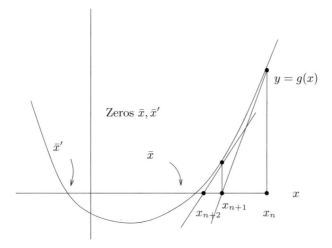

Figure 1.1: Newton's algorithm.

So

$$x_{n+1} = x_n - \frac{g(x_n)}{g'(x_n)} \equiv f(x_n),$$

where

$$f(x) \equiv x - \frac{g(x)}{g'(x)}.$$

The above iterations can encounter difficulty, for example, when:

(1) At $\bar{x}$, where $g(\bar{x}) = 0$, we also have $g'(\bar{x}) = 0$;

(2) The iterates x_n converge to a different (undesirable solution) $\bar{x}'$ instead of $\bar{x}$;

(3) The iterates x_j jump between two values x_n and x_{n+1}, such as what Fig. 1.2 shows in the following.

If any of the above happens, we have:

$$\lim_{n \to \infty} x_n \neq \bar{x}, \text{ for the desired solution } \bar{x}.$$

From now on, for any real-valued function f, we will use f^n to denote the n-th iterate of f defined by

$$f^n(x) = \underbrace{f(f(f \cdots (f(x)))) \cdots)}_{n\text{-times}} = \underbrace{f \circ f \circ f \cdots \circ f}_{n\text{-times}}(x),$$

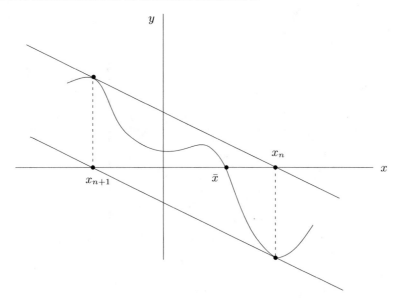

Figure 1.2: Newton's algorithm becomes stagnant at x_n and x_{n+1}.

if each $f^j(x)$ lies in the domain of definition of f for $j = 1, 2, \ldots, n-1$.

Exercise 1.2   Consider the iteration of the quadratic map:

$$\begin{cases} x_{n+1} = f_\mu(x_n); \quad f_\mu(x) = \mu x(1-x), \\ x_0 \in I \equiv [0, 1]. \end{cases}$$

(1) Choose $\mu = 3.2, 3.5, 3.55, 3.58, 3.65, 3.84$, and 3.94. For each given μ, plot the graphs:

$$y = f_\mu(x), y = f_\mu^2(x), y = f_\mu^3(x), y = f_\mu^4(x), y = f_\mu^5(x), y = f_\mu^{400}(x), x \in I,$$

where

$$f_\mu^2(x) = f_\mu(f_\mu(x)); \quad f_\mu^3(x) = f_\mu(f_\mu(f_\mu(x))), \text{ etc.}$$

(2) Let μ begin from $\mu = 2.9$ and increase μ to $\mu = 4$ with increment $\Delta\mu = 0.01$, with μ as the horizontal axis. For each μ, choose:

$$x_0 = \frac{k}{100}; \quad k = 1, 2, 3, \ldots, 99.$$

Plot $f_\mu^{400}(x_0)$ (i.e., a dot) for these values of x_0 on the vertical axis.   □

Example 1.3   The quadratic map f_μ as defined in Exercise 1.2 and shown in Fig. 1.3 is an example of a *unimodal map*. A map $f: I \equiv [a, b] \to I$ is said to be unimodal if it satisfies

$$f(a) = f(b) = a,$$

and f has a unique critical point $\bar{c}:\ a < \bar{c} < b$. The quadratic map f_μ is very representative of the dynamical behavior of unimodal maps.   □

Let f be a continuous function such that $f: I \to I$ on a closed interval I.
   A point $\bar{x}$ is said to be a *fixed point* of a map $y = f(x)$ if

$$\bar{x} = f(\bar{x}). \tag{1.6}$$

The set of all fixed points of f is denoted as Fix(f). A point $\bar{x}$ is said to be a *periodic point* with *prime period* k, if

$$\bar{x} = f^k(\bar{x}), \tag{1.7}$$

and k is the smallest positive integer to satisfy (1.7). The set of all periodic points of prime period k of f is denoted as Per$_k(f)$, and that of all periodic points of f is denoted as Per(f). In the analysis of the iterations $x_{n+1} = f(x_n)$, *fixed points and periodic points play a critical role*.
   Look at the quadratic map $f_\mu(x)$ in Fig. 1.3.

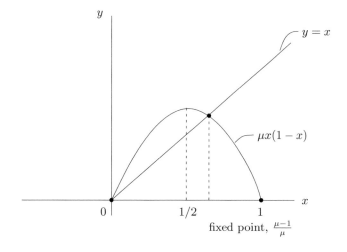

Figure 1.3: Graph of the Quadratic Map $y = f_\mu(x)$, $\mu = 2.7$. Its maximum always occurs at $x = 1/2$. It has a (trivial) fixed point at $x = 0$, and another fixed point at $x = \frac{\mu-1}{\mu}$.

The fixed point $\bar{x} = \frac{\mu-1}{\mu}$ can be *attracting* or *repelling*, as Fig. 1.4(a) and (b) have shown.

Definition 1.4   Let $\bar{x}$ be a periodic point of prime period p of a differentiable real-valued map $f:\ f^p(\bar{x}) = \bar{x}$. We say that $\bar{x}$ is attracting (resp., repelling) if

$$|(f^p)'(\bar{x})| < 1 \quad (\text{resp., } |(f^p)'(\bar{x})| > 1).$$

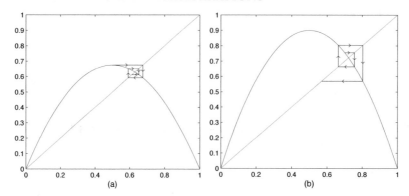

Figure 1.4: (a) The fixed point has slope $f'(\bar{x})$ such that $|f'(\bar{x})| < 1$. The iterates are attracted to the fixed point.
(b) The fixed point has slope $f'(\bar{x})$ such that $|f'(\bar{x})| > 1$. The iterates are moving away from $\bar{x}$.

A periodic point $\bar{x}$ of prime period n is said to be *hyperbolic* if $|(f^p)'(\bar{x})| \neq 1$, i.e., $\bar{x}$ must be either attracting or repelling.   □

   We present a few fundamental theorems.

Theorem 1.5   (Brouwer's Fixed Point Theorem) Let $I = [a, b]$, and let f be a continuous function on I

$$\text{such that either (i) } f(I) \subseteq I \text{ or (ii) } f(I) \supseteq I.$$

Then f has at least one fixed point on I.

Proof. Consider (i) first:
   Define: $g(x) = x - f(x)$. Because $f(I) \subseteq I$, i.e., $f([a, b]) \subseteq [a, b]$, so $f(a) \in [a, b]$, and $f(b) \in [a, b]$, thus $a \leq f(a) \leq b, a \leq f(b) \leq b$. Then

$$g(a) = a - f(a) \leq 0,$$
$$g(b) = b - f(b) \geq 0.$$

If equality does not hold in either of the two relations above, then

$$g(a) < 0; \quad g(b) > 0.$$

By the Intermediate Value Theorem, we get

$$g(c) = 0 \quad \text{for some} \quad c: a < c < b.$$

Thus, c is a fixed point of f. Next consider (ii):

$$f(I) \supseteq I, \quad [a, b] \subseteq f([a, b]).$$

Therefore, there exist $x_1, x_2 \in [a, b]$ such that

$$f(x_1) \leq a < b \leq f(x_2).$$

Again, define $g(x) = x - f(x)$. Then

$$g(x_1) = x_1 - f(x_1) \geq x_1 - a \geq 0,$$
$$g(x_2) = x_2 - f(x_2) \leq x_2 - b \leq 0.$$

Therefore, there exists a point $\tilde{x}$ in either $[x_1, x_2]$ or $[x_2, x_1]$, such that $g(\tilde{x}) = 0$. Thus, $\tilde{x}$ is a fixed point of f. □

Theorem 1.6   Let $f \colon I \to I$ be continuous where $I = [a, b]$, such that f' is also continuous, satisfying:

$$|f'(x)| < 1 \quad \text{on} \quad I.$$

Then f has a unique fixed point on I.

Proof. The existence of a fixed point has been proved in Theorem 1.5, so we need only prove uniqueness. Suppose both x_0 and y_0 are fixed points of f:

$$x_0 = f(x_0),$$
$$y_0 = f(y_0).$$

Then

$$\frac{f(y_0) - f(x_0)}{y_0 - x_0} = \frac{y_0 - x_0}{y_0 - x_0} = 1 = f'(c)$$

by the Mean Value Theorem, for some $c \colon x_0 < c < y_0$. But

$$|f'(c)| < 1.$$

This is a contradiction. □

1.2   THE INVERSE AND IMPLICIT FUNCTION THEOREMS

From now on, we denote vectors and vector-valued functions by bold letters.

   We state without proof two theorems which will be useful in future discussions.

Theorem 1.7   (The Inverse Function Theorem) Let U and V be two open sets in $\mathbb{R}^N$ and $\boldsymbol{f}: U \to V$ is C^r for some $r \geq 1$. Assume that

   (i) $\boldsymbol{x}^0 \in U$, $\boldsymbol{y}^0 \in V$, and $\boldsymbol{f}(\boldsymbol{x}^0) = \boldsymbol{y}^0$;

   (ii) $\nabla \boldsymbol{f}(\boldsymbol{x})|_{\boldsymbol{x}=\boldsymbol{x}^0}$ is nonsingular, where

$$
\nabla \boldsymbol{f}(\boldsymbol{x}) = \begin{bmatrix} \dfrac{\partial f_1(\boldsymbol{x})}{\partial x_1} & \cdots & \dfrac{\partial f_1(\boldsymbol{x})}{\partial x_N} \\ \dfrac{\partial f_2(\boldsymbol{x})}{\partial x_1} & \cdots & \dfrac{\partial f_2(\boldsymbol{x})}{\partial x_N} \\ \vdots & & \vdots \\ \dfrac{\partial f_N(\boldsymbol{x})}{\partial x_1} & \cdots & \dfrac{\partial f_N(\boldsymbol{x})}{\partial x_N} \end{bmatrix}.
$$

Then there exists an open neighborhood $N(\boldsymbol{x}^0) \subseteq U$ of $\boldsymbol{x}^0$ and an open neighborhood $N(\boldsymbol{y}^0) \subseteq V$ of $\boldsymbol{y}^0$ and a C^r-map $\boldsymbol{g}$:

$$
\boldsymbol{g}: N(\boldsymbol{y}^0) \longrightarrow N(\boldsymbol{x}^0),
$$

such that

$$
\boldsymbol{f}(\boldsymbol{g}(\boldsymbol{y})) = \boldsymbol{y},
$$

i.e., $\boldsymbol{g}$ is a local inverse of $\boldsymbol{f}$. $\square$

Theorem 1.8   (The Implicit Function Theorem) Let

$$
\left. \begin{array}{l} f_1(x_1, \ldots, x_m, y_1, \ldots, y_n) = 0, \\ f_2(x_1, \ldots, x_m, y_1, \ldots, y_n) = 0, \\ \quad \vdots \qquad\qquad\qquad \vdots \\ f_n(x_1, \ldots, x_m, y_1, \ldots, y_n) = 0 \end{array} \right\} \tag{1.8}
$$

be satisfied for all $\boldsymbol{x} = (x_1, \ldots, x_m) \in U$ and $\boldsymbol{y} = (y_1, \ldots, y_n) \in V$, where U and V are open sets in, respectively, $\mathbb{R}^m$ and $\mathbb{R}^N$, and

$$
f_i: U \times V \to \mathbb{R} \text{ is } C^r, \text{ for some } r \geq 1, \text{ for all } i = 1, 2, \ldots, n.
$$

Assume that for $\boldsymbol{x}^0 = (x_1^0, x_2^0, \ldots, x_m^0) \in U$ and $\boldsymbol{y}^0 = (y_1^0, y_2^0, \ldots, y_n^0) \in V$,

$$f_i(\boldsymbol{x}^0, \boldsymbol{y}^0) = 0 \quad \text{for} \quad i = 1, 2, \ldots, n,$$

$$[\nabla_y f_i(\boldsymbol{x}^0, \boldsymbol{y}^0)] = \begin{bmatrix} \dfrac{\partial f_1}{\partial y_1} & \cdots & \dfrac{\partial f_1}{\partial y_n} \\ \vdots & & \vdots \\ \dfrac{\partial f_n}{\partial y_1} & \cdots & \dfrac{\partial f_n}{\partial y_n} \end{bmatrix}_{\substack{\text{at } \boldsymbol{x}=\boldsymbol{x}^0 \\ \boldsymbol{y}=\boldsymbol{y}^0}} \qquad \text{is nonsingular.}$$

Then there exist an open neighborhood $N(\boldsymbol{x}^0) \subseteq U$ of $\boldsymbol{x}^0$ and an open neighborhood $N(\boldsymbol{y}^0) \subseteq V$ of $\boldsymbol{y}^0$, and a C^r-map $\boldsymbol{g} \colon N(\boldsymbol{x}^0) \longrightarrow N(\boldsymbol{y}^0)$, such that $\boldsymbol{y}^0 = \boldsymbol{g}(\boldsymbol{x}^0)$ and

$$f_i(\boldsymbol{x}, \boldsymbol{g}(\boldsymbol{y})) = 0, \quad \forall \boldsymbol{x} \in N(\boldsymbol{x}^0), \forall \boldsymbol{y} \in N(\boldsymbol{y}^0), \qquad i = 1, 2, \ldots, n,$$

i.e., locally, $\boldsymbol{y}$ is solvable in terms of $\boldsymbol{x}$ by $\boldsymbol{y} = \boldsymbol{g}(\boldsymbol{x})$. $\qquad \square$

If we write equations in (1.8) as

$$\boldsymbol{F}(\boldsymbol{x}, \boldsymbol{y}) = (f_1(\boldsymbol{x}, \boldsymbol{y}), f_2(\boldsymbol{x}, \boldsymbol{y}), \ldots, f_n(\boldsymbol{x}, \boldsymbol{y})) = \boldsymbol{0},$$

then taking the differential around $\boldsymbol{x} = \boldsymbol{x}^0$ and $\boldsymbol{y} = \boldsymbol{y}^0$, we have

$$\nabla_x \boldsymbol{F} \cdot d\boldsymbol{x} + \nabla_y \boldsymbol{F} \cdot d\boldsymbol{y} = \boldsymbol{0},$$

where $d\boldsymbol{x} \approx \boldsymbol{x} - \boldsymbol{x}^0$ and $d\boldsymbol{y} \approx \boldsymbol{y} - \boldsymbol{y}^0$. Thus,

$$\nabla_x \boldsymbol{F} \cdot (\boldsymbol{x} - \boldsymbol{x}^0) + \nabla_y \boldsymbol{F} \cdot (\boldsymbol{y} - \boldsymbol{y}^0) = \boldsymbol{0}.$$

An approximate solution of $\boldsymbol{y}$ in terms of $\boldsymbol{x}$ near $\boldsymbol{y} = \boldsymbol{y}_0$ is

$$\boldsymbol{y} \approx \boldsymbol{y}^0 + [\nabla_y \boldsymbol{F}(\boldsymbol{x}^0, \boldsymbol{y}^0)]^{-1} \nabla_x \boldsymbol{F} \cdot (\boldsymbol{x} - \boldsymbol{x}_0).$$

This explains intuitively why the invertibility of $\nabla_y \boldsymbol{F}(\boldsymbol{x}^0, \boldsymbol{y}^0)$ is useful.

The Implicit Function Theorem can be proved using the Inverse Function Theorem, but the proofs of both theorems can be found in most advanced calculus books so we omit them here.

Example 1.9   Consider the relation

$$f(x, y) = ax^2 + bx + c + y = 0. \tag{1.9}$$

We have

$$\frac{\partial f}{\partial x} = 2ax + b \neq 0, \quad \text{if} \quad x \neq -\frac{b}{2a};$$

$$\frac{\partial f}{\partial y} = 1 \neq 0.$$

Thus, y is always solvable in terms of x:

$$y = -(ax^2 + bx + c).$$

Here, we actually see that if $x = x^0 = -\frac{b}{2a}$, then x is *not uniquely solvable* in terms of y in a neighborhood of $x^0 = -\frac{b}{2a}$ because by the quadratic formula applied to (1.9), we have

$$x = \frac{-b \pm \sqrt{b^2 - 4a(c + y)}}{2a} = -\frac{b}{2a} \pm \frac{\sqrt{b^2 - 4a(c + y)}}{2a},$$

i.e., x is not unique.

On the other hand, if $x^0 \neq -\frac{b}{2a}$, then in a neighborhood of x^0, x is uniquely solvable in terms of y. For example, for $a = b = c = 1$, with $x^0 = 2$ and $y^0 = -7$,

$$x^0 = 2 \neq -\frac{b}{2a} = -\frac{1}{2}.$$

The (unique) solution of x in terms of y in a neighborhood of x is thus

$$x = \frac{-b + \sqrt{b^2 - 4a(c + y)}}{2a} = \frac{-1 + \sqrt{1 - 4(1 + y)}}{2}.$$

We discard the branch

$$x = \frac{-b - \sqrt{b^2 - 4a(c + y)}}{2a} = \frac{-1 - \sqrt{1 - 4(1 + y)}}{2}$$

because it doesn't satisfy

$$x^0 = 2 = \frac{-1 + \sqrt{1 - 4(1 - 7)}}{2} = \frac{-1 + 5}{2}. \qquad \square$$

Exercise 1.10   Assume that $a, b, c, d \in \mathbb{R}$, and $a \neq 0$. Consider the relation

$$f(x, y) = ax^3 + bx^2 + cx + d - y = 0, \qquad x, y \in \mathbb{R}$$

(i) Discuss the local solvability of real solutions x for given y by using the implicit function theorem.

(ii) Under what conditions does the function

$$y = g(x) = ax^3 + bx^2 + cx + d$$

have a local inverse? A global inverse? $\qquad \square$

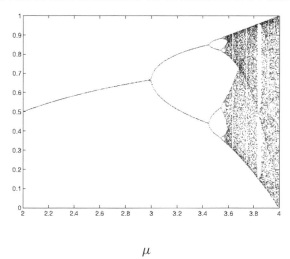

$$\mu$$

Figure 1.5: The orbit diagram of $f(\mu, x) = f_\mu(x) = \mu x(1 - x)$.

1.3   VISUALIZING FROM THE GRAPHICS OF ITERATIONS OF THE QUADRATIC MAP

In the next few pages, we discuss the computer graphics from the previous Exercise 1.2.

This type of graphics in Fig. 1.5 is called an *orbit diagram*. Note that the first period doubling happens at $\mu_0 = 3$. Then the second and third happen, respectively, at $\mu_1 \approx 3.45$ and $\mu_2 \approx 3.542$, and more period doublings happen in a cascade. We have

$$\frac{\mu_1 - \mu_0}{\mu_2 - \mu_1} \approx \frac{3.45 - 3}{3.542 - 3.45} = \frac{0.45}{0.092} \approx 4.8913\ldots.$$

It has been found that for *any period-doubling cascade*,

$$\lim_{n \to \infty} \frac{\mu_n - \mu_{n-1}}{\mu_{n+1} - \mu_n} = 4.669202\ldots.$$

This number, a *universal constant* due to M. Feigenbaum, is called the Feigenbaum constant. Also, note that there is a "window" area near $\mu = 3.84$ in Fig. 1.5.

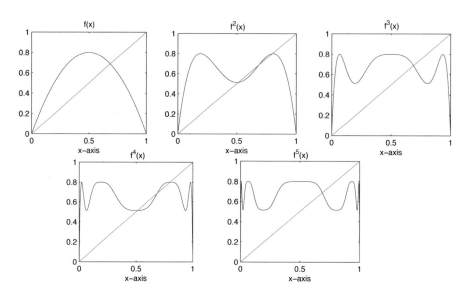

Figure 1.6: The graphics of $f_\mu(x)$, $f_\mu^2(x)$, $f_\mu^3(x)$, $f_\mu^4(x)$ and $f_\mu^5(x)$, where $\mu = 3.2$. Note that the intersections of the curves with the diagonal line $y = x$ represent either a fixed point or a periodic point.

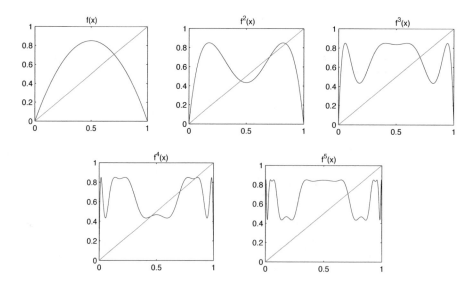

Figure 1.7: The graphics of $f_\mu(x)$, $f_\mu^2(x)$, $f_\mu^3(x)$, $f_\mu^4(x)$ and $f_\mu^5(x)$, where $\mu = 3.40$.

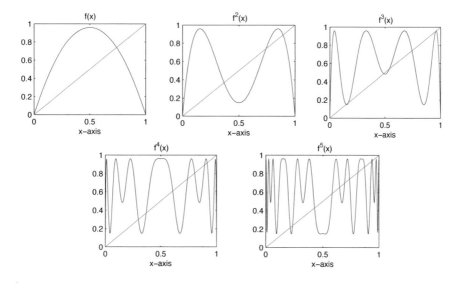

Figure 1.8: The graphics of $f_\mu(x)$, $f_\mu^2(x)$, $f_\mu^3(x)$, $f_\mu^4(x)$ and $f_\mu^5(x)$ where $\mu = 3.84$. We see that the curves have become more oscillatory (in comparison with those in Fig. 1.7). They intersect with the diagonal line $y = x$ at more points, implying that there are more periodic points. Note that $f_\mu^3(x)$ intersects with $y = x$ at P_1, P_2, ..., P_5 and P_6 (in addition to the fixed point $x = 0$). Each point P_i, $i = 1, 2, \ldots, 6$, has period 3. *If a continuous map has period 3, then it has period n for any n = 1, 2, 3, 4, ...*

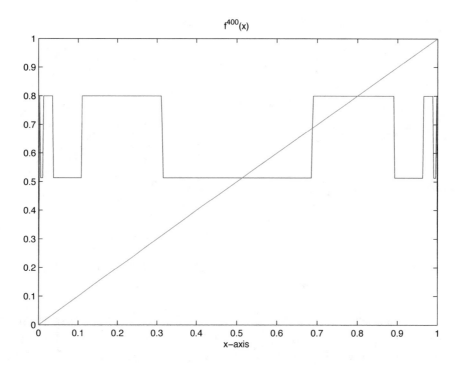

Figure 1.9: The graph of $f_\mu^{400}(x)$, where $\mu = 3.2$. It looks like a step function. The two horizontal levels correspond to the period-2 bifurcation curves in Fig. 1.5. Question: In the x-ranges close to $x = 0$ and $x = 1$, how oscillatory is the curve?

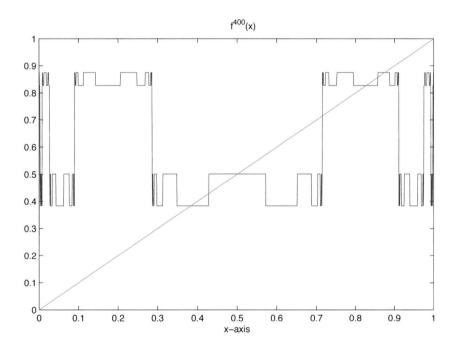

Figure 1.10: The graph of $f_\mu^{400}(x)$, where $\mu = 3.5$. It again looks like a step function, but with four horizontal levels.

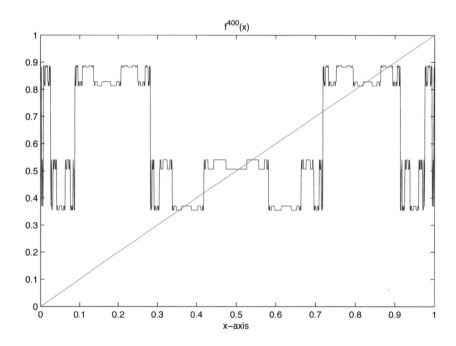

Figure 1.11: The graph of $f_\mu^{400}(x)$, with $\mu = 3.55$. This curve actually has *eight* horizontal levels.

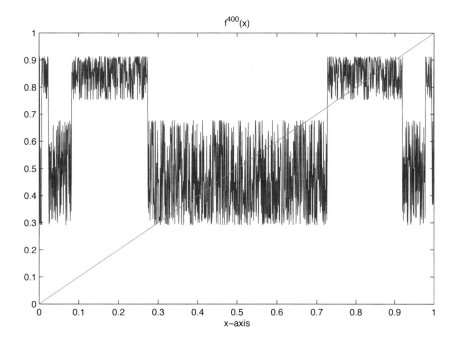

Figure 1.12: The graph of $f_\mu^{400}(x)$, $\mu = 3.65$. This value of μ is already in the chaotic regime. The curve has exhibited highly oscillatory behavior.

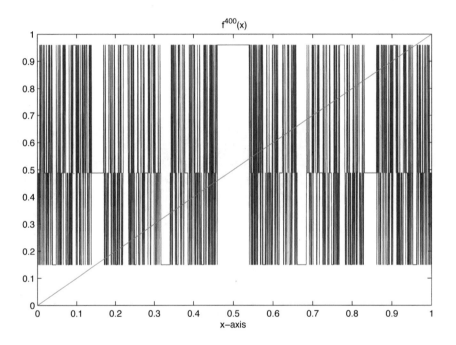

Figure 1.13: The graph of $f_\mu^{400}(x)$, $\mu = 3.84$. Note that this value of μ corresponds to the "window" area in Fig. 1.5. The curve is highly oscillatory, but it appears to take only three horizontal values.

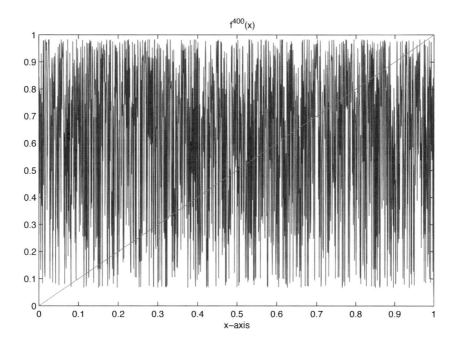

Figure 1.14: The graph of $f_\mu^{400}(x)$, with $\mu = 3.93$. This value of μ is also in the "chaotic regime".

Visualization from these graphics, Figs. 1.5–1.14, will provide inspirations for the study of the oscillatory behaviors related to chaos in this book.

Exercise 1.11   Define

$$y = f_\mu(x) = \mu \sin \pi x, \qquad x \in I = [0, 1].$$

(1) Vary $\mu \in [0, 1]$, plot the graphics of f_μ, f_μ^2, f_μ^3, ..., f_μ^{10}.

(2) Plot the orbit diagrams of $f_\mu(x)$.

(3) Describe what happens if $\mu > 1$.   □

Exercise 1.12   Pick your arbitrary favorite continuous function of the form

$$y = f(\mu, x), \qquad x \in [0, 1],$$

such that $f(\mu, \cdot)$ maps I into I for the parameter μ lying within a certain range. Plot the graphics of the various iterates of $f(\mu, \cdot)$ as well as orbit diagrams of $f(\mu, \cdot)$.   □

NOTES FOR CHAPTER 1

An *interval map* is formed by the 1-step scalar *equation of iteration* $x_{n+1} = f(x_n)$ for a continuous map f. Thus, it constitutes the simplest model for iterations. For example, Newton's method for finding roots of a nonlinear equation, and the time-marching of a 1-step explicit Euler finite-difference scheme for a first order scalar ordinary differential equation, can both result in an interval map. Interestingly, even for partial differential equations such as the nonlinear initial-boundary value problem of the wave equation in Appendix B, interval maps have found good applications.

Most of the textbooks on dynamical systems use the quadratic (or logistic) map (1.4) as a standard example to illustrate many peculiar, amazing behaviors of the iterates of the quadratic map. In fact, those iterates manifest strong chaotic phenomena which facilitates the understanding of what chaos is for pedagogical purposes. The focus of the first five chapters of this book is almost exclusively on interval maps.

The books by Devaney [20] and Robinson [58] contain excellent treatments of interval maps. The monograph by Block and Coppel [7] contains a more detailed account and further references about interval maps.

CHAPTER 2

Total Variations of Iterates of Maps

2.1 THE USE OF TOTAL VARIATIONS AS A MEASURE OF CHAOS

Let $f: I = [a, b] \to \mathbb{R}$ be a given function; f is not necessarily continuous. A *partition* of I is defined as

$$P = \{x_0, x_1, \ldots, x_n \mid x_j \in I, \text{ for } j = 0, 1, \ldots, n; a = x_0 < x_1 < x_2 < \cdots < x_n = b\},$$

which is an arbitrary finite collection of (ordered) points on I. Define

$$V_I(f) = \text{the total variation of } f \text{ on } I$$

$$= \sup_{\text{all } P} \left\{ \sum_{i=1}^{n} |f(x_i) - f(x_{i-1})| \;\Big|\; x_i \in P \right\}. \tag{2.1}$$

If f is continuous on I, and f has finitely many maxima and minima on I, such as indicated in Fig. 2.1 below. Then it is easy to see that

$$V_I(f) = |f(\tilde{x}_1) - f(\tilde{x}_0)| + |f(\tilde{x}_2) - f(\tilde{x}_1)| + \cdots + |f(\tilde{x}_n) - f(\tilde{x}_{n-1})|,$$

where each interval $[\tilde{x}_i, \tilde{x}_{i+1}]$ is a maximal interval where f is either increasing or decreasing.

Let I_1 and I_2 be two closed intervals, and let f be continuous such that

$$f(I_1) \supseteq I_2.$$

We write the above as

$$I_1 \xrightarrow{f} I_2 \quad \text{or} \quad I_1 \longrightarrow I_2$$

and say that I_1 f-covers I_2.

Lemma 2.1 If $I_1 \xrightarrow{f} I_2$, then $V_{I_1}(f) \geq |I_2| \equiv$ length of I_2.

Proof. This follows easily from the observation of Fig. 2.2. □

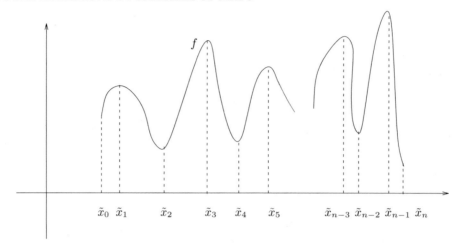

Figure 2.1: A continuous function with finitely many maxima and minima at $\tilde{x}_0, \tilde{x}_1, \ldots, \tilde{x}_n$.

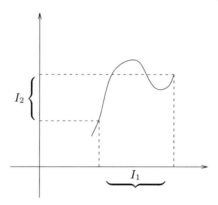

Figure 2.2: Interval I_1 f-covers I_2.

Lemma 2.2   Let $J_0, J_1, \ldots, J_{n-1}$ be bounded closed intervals such that they overlap at most at endpoints pairwise. Assume that $J_0 \longrightarrow J_1 \longrightarrow J_2 \longrightarrow \cdots \longrightarrow J_{n-1} \longrightarrow J_n \equiv J_0$ holds. Then

(i) there exists a fixed point x_0 of f^n: $f^n(x_0) = x_0$, such that $f^k(x_0) \in J_k$ for $k = 0, \ldots, n$;

(ii) Further, assume that the loop $J_0 \longrightarrow J_1 \longrightarrow \cdots \longrightarrow J_n$ is not a repetition of a shortened repetitive loop m where $mk = n$ for some integer $k > 0$. If the point x_0 in (i) is in the <u>interior</u> of J_0, then x_0 has prime period n.

Proof. Use mathematical induction.   □

Theorem 2.3   Let I be a closed interval and $f: I \to I$ be continuous. Assume that f has two fixed points on I and a pair of period-2 points on I. Then

$$\lim_{n \to \infty} V_I(f^n) = \infty.$$

Proof. Let the two fixed points be x_0 and x_1:

$$f(x_0) = x_0 \quad \text{and} \quad f(x_1) = x_1. \tag{2.2}$$

Let the two period-2 points be p_1 and p_2:

$$f(p_1) = p_2, \qquad f(p_2) = p_1. \tag{2.3}$$

Then there are three possibilities:

(i) $p_1 < x_0 < x_1 < p_2$;   (2.4)
(ii) $x_0 < p_1 < p_2$;
(iii) $p_1 < p_2 < x_1$.

We consider case (i) only. Cases (ii) and (iii) can be treated in a similar way.

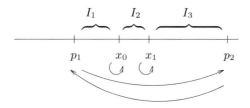

Figure 2.3: The points and intervals corresponding to (2.2), (2.3) and (2.4).

From Fig. 2.3, we see that we have

$$\left.\begin{aligned}
f(I_1) &\supseteq I_2 \cup I_3, && \text{i.e.,} && I_1 \longrightarrow I_2 \cup I_3, \\
f(I_2) &\supseteq I_2, && \text{i.e.,} && I_2 \longrightarrow I_2, \\
f(I_3) &\supseteq I_1 \cup I_2, && \text{i.e.,} && I_3 \longrightarrow I_1 \cup I_2.
\end{aligned}\right\} \tag{2.5}$$

Therefore, we have the covering diagram

It is easy to verify by mathematical induction that the following statement is true:

"For each n, I_1 contains $n + 1$ subintervals $\{I_{1,1}^{(n)}; I_{1,2}^{(n)}, \dots, I_{1,n+1}^{(n)}\}$ such that

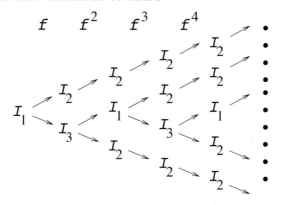

Figure 2.4: The covering of intervals according to (2.5).

$$I_{1,j}^{(n)} \subseteq I_1, \qquad I_{1,j_1}^{(n)} \cap I_{1,j_2}^{(n)} \quad \text{has empty interior if } j_1 \neq j_2,$$
$$f^n(I_{1,j}^{(n)}) \supseteq I_k \quad \text{for some} \quad k \in \{1, 2, 3\}\text{''}.$$

Therefore,

$$V_I(f^n) \geq V_{I_1}(f^n) \geq \sum_{j=1}^{n+1} V_{I_{1,j}^{(n)}}(f^n)$$

$$\geq (n + 1) \cdot \min\{|I_1|, |I_2|, |I_3|\} \to \infty \text{ as } n \to \infty. \tag{2.6}$$

$\square$

Exercise 2.4   Prove that for the quadratic map

$$f_\mu(x) = \mu x(1 - x), \qquad x \in I = [0, 1],$$

if $\mu > 3$, then f_μ has two fixed points and at least a pair of period-2 points. Therefore, the assumptions of Theorem 2.3 are satisfied and

$$\lim_{n \to \infty} V_I(f_\mu^n) = \infty \quad \text{for all} \quad \mu: \ 3 < \mu < 4.$$

$\square$

Theorem 2.5   Let I be a bounded interval and let $f : I \to I$ be continuous such that f has a period-3 orbit $\{p_1, p_2, p_3\}$ satisfying $f(p_1) = p_2$, $f(p_2) = p_3$ and $f(p_3) = p_1$. Then

$$\lim_{n \to \infty} V_I(f^n) \geq K e^{\alpha n} \quad \text{for some} \quad K, \alpha > 0, \tag{2.7}$$

i.e., the total variation of f^n grows exponentially with n.

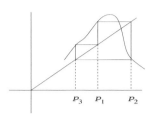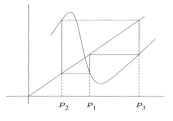

Figure 2.5: Two period-3 orbits satisfying $f(p_1) = p_2$, $f(p_2) = p_3$, $f(p_3) = p_1$.

Proof. To help our visualization, we draw the graphics of a period-3 orbit in Fig. 2.5.
We have two possibilities:

$$\text{(i) } p_2 < p_1 < p_3, \quad \text{or} \quad \text{(ii) } p_3 < p_1 < p_2. \tag{2.8}$$

Here we treat only case (i). Define

$$I_1 = [p_1, p_2], \qquad I_2 = [p_3, p_1].$$

Then

$$\left.\begin{array}{ll} f(I_1) \supseteq I_1 \cup I_2, & \text{i.e., } \; I_1 \longrightarrow I_1 \cup I_2, \\ f(I_2) \supseteq I_1, & \text{i.e., } \; I_2 \longrightarrow I_1. \end{array}\right\} \tag{2.9}$$

Thus, we have the covering diagram in Fig. 2.6.
For each n, one can prove by mathematical induction that if the $(n + 1)$th column (after mapping by f^n) contains a_n subintervals of I_1 or I_2, then the following relation is satisfied:

$$\begin{cases} a_{n+1} = a_n + a_{n-1}, & \text{for } n = 2, 3, 4, \ldots, \\ a_1 = 2, \quad a_2 = 3. \end{cases} \tag{2.10}$$

An exact solution to the recurrence relation (2.10) can be determined as follows. Assume that a solution of $a_{n+1} = a_n + a_{n-1}$ can be written in the form

$$a_k = cx^k, \quad \text{for} \quad k = 1, 2, \ldots. \tag{2.11}$$

Then substituting (2.11) into the first equation of (2.10) gives

$$cx^{n+1} = cx^n + cx^{n-1},$$
$$x^2 - x - 1 = 0,$$
$$x = \frac{1 \pm \sqrt{5}}{2}.$$

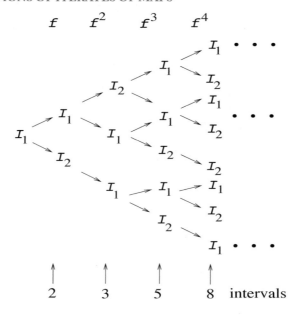

Figure 2.6: The covering diagram satisfying (2.9).

Therefore, we write the solution of (2.10) as

$$
a_n = c_1 \left(\frac{1 + \sqrt{5}}{2} \right)^n + c_2 \left(\frac{1 - \sqrt{5}}{2} \right)^n, \qquad n = 1, 2, \ldots,
$$

$$
a_1 = 2 = c_1 \left(\frac{1 + \sqrt{5}}{2} \right) + c_2 \left(\frac{1 - \sqrt{5}}{2} \right),
$$

$$
a_2 = 3 = c_1 \left(\frac{1 + \sqrt{5}}{2} \right)^2 + c_2 \left(\frac{1 - \sqrt{5}}{2} \right)^2,
$$

and obtain

$$
c_1 = \frac{5 + 3\sqrt{5}}{10}, \qquad c_2 = \frac{5 - 3\sqrt{5}}{10}.
$$

It is easy to show that

$$
a_n = c_1 \left(\frac{1 + \sqrt{5}}{2} \right)^n + c_2 \left(\frac{1 - \sqrt{5}}{2} \right)^n
$$

$$
\geq k_0 \left(\frac{1 + \sqrt{5}}{2} \right)^n
$$

for some $k_0 > 0$, for all $n = 1, 2, \ldots$. Therefore, using the same arguments as in the proof of Theorem 2.3, we have

$$V_I(f^n) \geq V_{I_1}(f^n) \geq k_0 \left(\frac{1 + \sqrt{5}}{2} \right)^n \cdot \min\{|I_1|, |I_2|\} \equiv k e^{\alpha n}, \text{ for } n = 1, 2, \ldots$$

where

$$k \equiv k_0 \cdot \min\{|I_1|, |I_2|\} \quad \text{and} \quad \alpha \equiv \ln \left(\frac{1 + \sqrt{5}}{2} \right) > 0. \tag{2.12}$$

$\square$

Using the same arguments as in the proof of Theorem 2.5, we can also establish the following.

Theorem 2.6   Let I be a bounded closed interval and $f : I \to I$ be continuous. Assume that $I_1, I_2, \ldots, I_n$ are closed subintervals of I which overlap at most at endpoints, and the covering relation

$$I_1 \longrightarrow I_2 \longrightarrow I_3 \longrightarrow \cdots \longrightarrow I_n \longrightarrow I_1 \cup I_j, \text{ for some } j \neq 1. \tag{2.13}$$

Then for some $K > 0$ and $\alpha > 0$,

$$V_I(f^n) \geq K e^{\alpha n} \longrightarrow \infty, \quad \text{as} \quad n \to \infty. \quad \square \tag{2.14}$$

Theorem 2.6 motivates us to give the following definition of chaos, the first one of such definitions in this book.

Definition 2.7   Let $f : I \to I$ be an interval map such that there exist $K > 0, \alpha > 0$ such that (2.14) holds. We say that f is *chaotic in the sense of exponential growth of total variations of iterates*.$\square$

In Chapter 9, such a map f will also be said to have *rapid fluctuations of dimension 1*.

Corollary 2.8   Let $f : I \to I$ be an interval map satisfying (2.13) in the assumption of Theorem 2.6. Then f is chaotic in the sense of exponential growth of variations of iterates. $\square$

NOTES FOR CHAPTER 2

The total variation of a scalar-valued function on an interval provides a numerical measure of how strong the oscillatory behavior that function has, when the interval is finite. This chapter is based on G. Chen, T. Huang and Y. Huang [17]. It shows that the total variations of iterates of a given map can be bounded, of polynomial growth, and of exponential growth. Only the case of exponential growth of total variations of iterates is classified as chaos (while the case of polynomial growth is associated with the existence of periodic points). This offers a *global approach* to the study of Chaotic maps.

This chapter will pave the way for the study of chaotic behavior in terms of total variations in higher and fractional dimensions in Chapters 8 and 9.

CHAPTER 3

Ordering among Periods: The Sharkovski Theorem

One of the most beautiful theorems in the theory of dynamical systems is the Sharkovski Theorem. An interval map may have many different periodic points with seemingly unrelated periodicities. What is unexpected and, in fact, amazing is that those periodicities are actually oredered in a certain way, called the Sharkovski ordering. The "top chain of the ordering" consists of all odd integers, with the number 3 at the zenith, and the "bottom chain of the ordering" consists of all decreasing powers of 2, with the number 1 at the nadir.

Here we give the statement of the theorem and provide a sketch of ideas of the proof. We introduce the Sharkovski ordering on the set of all positive integers. The ordering is arranged as in Fig. 3.1.

$$
\begin{array}{cccccccc}
3 & 2 \cdot 3 & 2^2 \cdot 3 & \vdots & 2^n \cdot 3 & \vdots & \vdots \\
\triangledown & \triangledown & \triangledown & & \triangledown & & 2^{m+1} \\
5 & 2 \cdot 5 & 2^2 \cdot 5 & & 2^n \cdot 5 & & \triangledown \\
\triangledown & \triangledown & \triangledown & & \triangledown & & 2^m \\
7 & 2 \cdot 7 & 2^2 \cdot 7 & & 2^n \cdot 7 & & \triangledown \\
\triangledown & \triangledown & \triangledown & & \triangledown & & 2^{m-1} \\
9 & 2 \cdot 9 & 2^2 \cdot 9 & & 2^n \cdot 9 & & \triangledown \\
\triangledown & \triangledown & \triangledown & & \triangledown & & \vdots \\
\vdots & \vdots & \vdots & & \vdots & & \triangledown \\
\triangledown & \triangledown & \triangledown & & \triangledown & & 2^3 \\
2n+1 & 2 \cdot (2n+1) & 2^2 \cdot (2n+1) & & 2^n \cdot (2m+1) & & \triangledown \\
\triangledown & \triangledown & \triangledown & & \triangledown & & 2^2 \\
2n+3 & 2 \cdot (2n+3) & 2^2 \cdot (2n+3) & & 2^n \cdot (2m+3) & & \triangledown \\
\triangledown & \triangledown & \triangledown & & \triangledown & & 2 \\
\vdots & \vdots & \vdots & & \vdots & & \triangledown \\
& & & & & & 1.
\end{array}
$$

Figure 3.1: The Sharkovski ordering.

Theorem 3.1   (Sharkovski's Theorem)   Let I be a bounded closed interval and $f : I \to I$ be continuous. Let $n \triangleright k$ in Sharkovski's ordering. If f has a (prime) period n orbit, then f also has a (prime) period k orbit. □

The following lemma is key in the proof of Sharkovski's Theorem.

Lemma 3.2   Let n be an odd integer. Let f have a periodic point of prime period n. Then there exists a periodic orbit $\{x_j \mid j = 1, 2, \ldots n;\ f(x_j) = x_{j+1}$ for $j = 1, 2, \ldots, n-1;\ f(x_n) = x_1\}$ of prime period n such that either

$$x_n < \cdots < x_5 < x_3 < x_1 < x_2 < x_4 < \cdots < x_{n-1} \tag{3.1}$$

or

$$x_{n-1} < \cdots < x_4 < x_2 < x_1 < x_3 < x_5 < \cdots < x_n. \tag{3.2}$$

□

The proof of Lemma 3.2 may be found in Robinson [58, pp. 67–69].

We now consider (3.1) only; (3.2) is a mirror image of (3.1) and can be treated similarly. We define subintervals $I_1, I_2, \ldots, I_n$ according to Fig. 3.2.

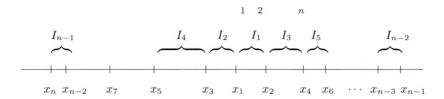

Figure 3.2: The subintervals $I_1, I_2, \ldots, I_n$ according to (3.1), where $I_{2j-1} = [x_{2j-2}, x_{2j}]$ and $I_{2j} = [x_{2j+1}, x_{2j-1}]$, for $j = 2, \ldots, (n-1)/2$.

We now look at the covering relations of intervals $I_1, I_2, \ldots, I_n$. From

$$f(x_1) = x_2, \qquad f(x_2) = x_3,$$

we have

$$f(I_1) \supseteq I_1 \cup I_2. \tag{3.3}$$

From

$$f(x_3) = x_4, \qquad f(x_1) = x_2,$$

we have

$$f(I_2) \supseteq I_3. \tag{3.4}$$

Similarly,

$$f(I_3) \supseteq I_4,\ f(I_4) \supseteq I_5, \ldots,\ f(I_{n-2}) \supseteq I_{n-1}. \tag{3.5}$$

However,

$$I_{n-1} = [x_n, x_{n-2}], \quad f(x_n) = x_1, \quad f(x_{n-2}) = x_{n-1},$$

and, therefore

$$f(I_{n-1}) \supseteq [x_1, x_{n-1}] = I_1 \cup I_3 \cup I_5 \cup \cdots \cup I_{n-2}. \tag{3.6}$$

Example 3.3 In Lemma 3.2, let $n = 7$. Then (3.3)–(3.6) above give us the following diagram in Fig. 3.3. $\qquad\square$

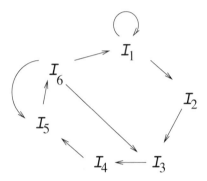

Figure 3.3: Covering relations for intervals $I_1, I_2, \ldots, I_6$ where $n = 7$ in Lemma 3.2. Note that I_1 covers both I_1 and I_2, while I_6 covers all the odd-numbered intervals I_1, I_3 and I_5.

For a general odd positive integer n, from (3.3)–(3.6), we can construct the graph in Fig. 3.4, called the *Stefan cycle*.

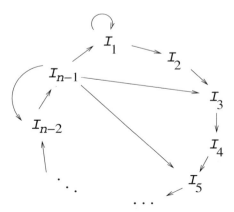

Figure 3.4: The Stefan cycle corresponding to (3.1).

Proposition 3.4   Assume that n is odd and k is a positive integer such that $n \triangleright k$ in Sharkovski's ordering. If f has a prime period n, then f also has a prime period k.

Proof. There are two possibilities: (i) k is even and $k < n$; and (ii) $k > n$ and k can be either even or odd.

Consider Case (i) first. We use the loop

$$I_{n-1} \longrightarrow I_{n-k} \longrightarrow I_{n-k+1} \longrightarrow \cdots \longrightarrow I_{n-2} \longrightarrow I_{n-1}. \tag{3.7}$$

$$\uparrow$$

$$\text{note: } n - k \text{ here is odd.}$$

Then by Lemma 2.2 in Chapter 2, there exists an $x_0 \in I_{n-1}$ with period k. The point x_0 cannot be an endpoint because the endpoints have period n. Therefore, x_0 has prime period k.

Next, we consider Case (ii). We use the following loop of length k:

$$I_1 \longrightarrow I_2 \longrightarrow \cdots \longrightarrow I_{n-1} \longrightarrow I_1 \longrightarrow \underbrace{I_1 \longrightarrow \cdots \longrightarrow I_1}_{k-n+1 \ I_1\text{'s}} \tag{3.8}$$

Thus, there exists an $x_0 \in I_1$ with $f^k(x_0) = x_0$. If x_0 is an endpoint of I_1, then x_0 has period n and, therefore, k is divisible by n, and $k \geq 2n \geq n + 3$. Either $x_0 = x_1$ or $x_0 = x_2$ is satisfied for x_1 and x_2 in (3.1). Thus,

(a) if $x_0 = x_1$, then $f^n(x_0) = x_0$ and $f^{n+2}(x_0) = f^{n+2}(x_1) = f^2(x_1) = x_3$;

(b) if $x_0 = x_2$, then $f^n(x_0) = x_0$ and $f^{n+2}(x_0) = f^{n+2}(x_2) = f^2(x_2) = x_4$.

In either case above, $f^{n+2}(x_0) \notin I_1$. This violates our choice of x_0 that it satisfies

$$x_0 \in I_1 \longrightarrow I_2 \longrightarrow I_3 \longrightarrow \cdots \longrightarrow I_{n-1} \longrightarrow I_1 \longrightarrow I_1$$
$$\downarrow \quad\quad \downarrow \quad\quad\quad\quad\quad\quad\quad\quad\quad\quad \downarrow \quad\quad \downarrow$$
$$f(x_0) \quad f^2(x_0) \quad\quad\quad\quad\quad\quad\quad f^{n-1}(x_0) \quad f^n(x_0)$$
$$\longrightarrow I_1 \longrightarrow I_1 \longrightarrow \cdots \longrightarrow I_1,$$
$$\downarrow \quad\quad \downarrow$$
$$f^{n+1}(x_0) \quad f^{n+2}(x_0)$$

from (3.8) because $k - n \geq n + 3$. □

For readers who are interested to see a simple, complete proof of the Sharkovski Theorem, we recommend Du [23, 24, 25].

NOTES FOR CHAPTER 3

A.N. Sharkovski (1936-) published his paper [62] in the Ukrainian Mathematical Journal in 1964. This paper was ahead of the time before iterations, chaos and nonlinear phenomena became fashionable. Also, the paper was written in Russian. Thus, Sharkovski's results went unnoticed for a decade.

The American mathematicians Tien-Yien Li and James A. Yorke published a famous paper entitled "Period three implies chaos" in the American Mathematical Monthly in 1975. They actually proved a special part of Sharkovski's result besides coining the term chaos. Li and Yorke attended a conference in East Berlin where they met Sharkovski. Although they could not converse in a common language, the meeting led to global recognition of Sharkovski's work.

Sharkovski's Theorem does not hold for multidimensional maps. For circle maps, rotation by one hundred twenty degrees is a map with period three. But it does not have any other periods.

Today, there exist several ways of proving Sharkovski's Theorem: by Stefan [66], Block, Guckenheimer, Misiurewicz and Young [8], Burkart [9], Ho and Morris [33], Ciesielski and Pogoda [19], and Du [23, 24], and others.

CHAPTER 4

Bifurcation Theorems for Maps

Bifurcation means "branching". It is a major nonlinear phenomenon. Bifurcation happens when one or several important system parameters change values in a transition process. After a bifurcation, the system's behavior changes. For example, new equilibrium states emerge, with a different behavior, especially that related to stability.

4.1 THE PERIOD-DOUBLING BIFURCATION THEOREM

Period doubling is an important *chaos route to*. We have seen from Fig. 1.4 that the (local) diagram looks like what is shown in Fig. 4.1.

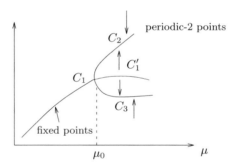

Figure 4.1: Period doubling and stability of bifurcated solutions.

In Fig. 4.1, the C_1 branch of fixed points loses its stability at $\mu = \mu_0$ and bifurcates into $C_2 \cup C_3$, which is a curve of period-2 points. We want to analyze this bifurcation. In performing the analysis, there are at least two difficulties involved:

(i) For the iteration $x_{n+1} = f_\mu(x_n) = f(\mu, x_n)$, any period-2 point $\bar{x}$ satisfies

$$\bar{x} = f_\mu^2(\bar{x}) = f(\mu, f(\mu, \bar{x})). \tag{4.1}$$

But if $\tilde{x}$ is a fixed point, $\tilde{x}$ will also satisfy (4.1). How do we pick out those period-2 points $\bar{x}$ which are not fixed points?

(ii) Assume that we can resolve (i) above. Can we also determine the stability (i.e., whether attracting, or repelling) of the period-2 points?

These will be answered in the following theorem.

Theorem   4.1   (Period Doubling Bifurcation Theorem)   Consider the map
$f(\mu, \cdot): I \rightarrow I$ where I is a closed interval and f is C^r for $r \geq 3$. Let the curve C represent a family of fixed points of $f(\mu, \cdot)$, where $C: x = x(\mu)$, and $x(\mu)$ satisfies

$$x(\mu) = f(\mu, x(\mu)).$$

Assume that at $\mu = \mu_0$,

(i) $$\left. \frac{\partial f(\mu, x)}{\partial x} \right|_{\substack{\mu=\mu_0 \\ x=x(\mu_0) \equiv x_0}} = -1,$$ (4.2)

(ii) $$\left[\frac{\partial^2 f(\mu, x)}{\partial \mu \partial x} + \frac{1}{2} \frac{\partial f(\mu, x)}{\partial \mu} \frac{\partial^2 f(\mu, x)}{\partial x^2} \right] \Bigg|_{\substack{\mu=\mu_0 \\ x=x_0}} = \alpha \neq 0,$$ (4.3)

then there is period-doubling bifurcation at $\mu = \mu_0$ and $x = x_0$, i.e., there exists a curve γ of the parameter μ in terms of x: $\mu = m(x)$ in a neighborhood of $\mu = \mu_0$ and $x = x_0$ such that $\mu_0 = m(x_0)$ and

$$x = f_\mu^2(x)|_{\mu=m(x)} = f(\mu, f(\mu, x))|_{\mu=m(x)}.$$ (4.4)

Further, assume that

$$\left\{ \frac{1}{3!} \frac{\partial^3 f(\mu, x)}{\partial x^3} + \left[\frac{1}{2!} \frac{\partial^2 f(\mu, x)}{\partial x^2} \right]^2 \right\} \Bigg|_{\substack{\mu=\mu_0 \\ x=x_0}} = \beta \neq 0.$$ (4.5)

Then the bifurcated period-2 points on the curve $\mu = m(x)$ are attracting if $\beta > 0$ and repelling if $\beta < 0$.   □

Proof. We follow the proof in [20]. The fixed points x of f_μ satisfy

$$x = f_\mu(x) = f(\mu, x),$$

and, thus,

$$F(\mu, x) \equiv f(\mu, x) - x = 0.$$ (4.6)

At $x = x(\mu_0) = x_0$ and $\mu = \mu_0$,

$$\frac{\partial}{\partial x} F(\mu, x)|_{\substack{\mu=\mu_0 \\ x=x_0}} = \left[\frac{\partial f(\mu, x)}{\partial x} - 1 \right] \Bigg|_{\substack{\mu=\mu_0 \\ x=x_0}} = -1 - 1 = -2 \neq 0.$$

Therefore, by Theorem 4.1, the Implicit Function Theorem, x is uniquely solvable (locally near $x = x_0$) in terms of μ:

$$x = x(\mu), \quad \text{such that} \quad x_0 = x(\mu_0).$$

This gives us the curve $C: x = x(\mu)$ of fixed points.

Next, we want to capture the bifurcated period-2 points near $x = x_0$ and $\mu = \mu_0$. These points satisfy (4.1). To simplify notation, let us define

$$g(\mu, y) = f(\mu, y + x(\mu)) - x(\mu). \tag{4.7}$$

Then it is easy to check that

$$\frac{\partial^j}{\partial y^j} g(\mu, y)\Big|_{y=0} = \frac{\partial^j}{\partial x^j} f(\mu, x)\Big|_{x=x(\mu)}, \quad \text{for} \quad j = 1, 2, 3, \dots. \tag{4.8}$$

This change of variable will give us plenty of convenience. We note that $y = y(\mu) \equiv 0$ becomes the curve of fixed points for the map g. Since $g(\mu, y)|_{y=0} = g(\mu, 0) = 0$, we have the Taylor expansion

$$g(\mu, y) = a_1(\mu)y + a_2(\mu)y^2 + a_3(\mu)y^3 + \mathcal{O}(|y|^4).$$

The period-2 points of f_μ now satisfy

$$\begin{aligned}
y = g_\mu^2(y) &= g(\mu, g(\mu, y)) \\
&= a_1(\mu)[a_1(\mu)y + a_2(\mu)y^2 + a_3(\mu)y^3 + \mathcal{O}(|y|^4)] \\
&\quad + a_2(\mu)[a_1(\mu)y + a_2(\mu)y^2 + a_3(\mu)y^3 + \mathcal{O}(|y|^4)]^2 \\
&\quad + a_3(\mu)[a_1(\mu)y + a_2(\mu)y^2 + a_3(\mu)y^3 + \mathcal{O}(|y|^4)]^3 + \mathcal{O}(|y|^4) \\
y &= a_1^2 y + (a_1 a_2 + a_1^2 a_2)y^2 + (a_1 a_3 + 2a_1 a_2^2 + a_1^3 a_3)y^3 + \mathcal{O}(|y|^4),
\end{aligned}$$

where in the above, we have omitted the dependence of a_1, a_2 and a_3 on μ. Since $y = y(\mu) = 0$ corresponds to the fixed points of f_μ, we don't want them. Therefore, define

$$M(\mu, y) = \frac{g_\mu^2(y) - y}{y} \quad \text{if} \quad y \neq 0. \tag{4.9}$$

This gives

$$M(\mu, y) = (a_1^2 - 1) + (a_1 a_2 + a_1^2 a_2)y + (a_1 a_3 + 2a_1 a_2^2 + a_1^3 a_3)y^2 + \mathcal{O}(|y|^3).$$

The above function $M(\mu, y)$ is obviously C^2 even for $y = 0$. Thus, we can extend the definition of $M(\mu, y)$ given in (4.9) by continuity even to $y = 0$. Now, note that period-2 points are determined by the equation $M(\mu, y) = 0$. We have

$$\begin{aligned}
\frac{\partial}{\partial \mu} M(\mu, y)\Big|_{\substack{\mu=\mu_0 \\ y=0}} &= 2a_1(\mu_0)a_1'(\mu_0) \\
&= 2\left[\frac{\partial g(\mu, y)}{\partial y}\Big|_{\substack{\mu=\mu_0 \\ y=0}}\right]\left[\frac{\partial^2 g(\mu, y)}{\partial \mu \partial y}\Big|_{\substack{\mu=\mu_0 \\ y=0}}\right] \\
&= 2\left[\frac{\partial f(\mu_0, x_0)}{\partial x}\right]\left[\frac{\partial^2 f(\mu_0, x_0)}{\partial \mu \partial x} + \frac{\partial^2 f(\mu_0, x_0)}{\partial x^2} x'(\mu_0)\right] \\
&= -2\alpha, \tag{4.10}
\end{aligned}$$

where we have utilized the fact that

$$
\begin{aligned}
x(\mu) &= f(\mu, x(\mu)), \\
x'(\mu) &= \frac{\partial f}{\partial \mu} + \frac{\partial f}{\partial x} x'(\mu), \\
x'(\mu) &= \frac{\partial f / \partial \mu}{1 - \frac{\partial f}{\partial x}}, \\
x'(\mu_0) &= \frac{\partial f(\mu_0, x_0) / \partial \mu}{1 - \frac{\partial f(\mu_0, x_0)}{\partial x}} = \frac{1}{2} \frac{\partial f(\mu_0, x_0)}{\partial \mu}.
\end{aligned}
$$

From (4.3) and (4.10), we thus have

$$
\frac{\partial}{\partial \mu} M(\mu, \dot{y}) \Big|_{\substack{\mu = \mu_0 \\ y = 0}} = -2\alpha \neq 0. \tag{4.11}
$$

On the other hand,

$$
\begin{aligned}
\frac{\partial}{\partial y} M(\mu, y) \Big|_{\substack{\mu = \mu_0 \\ y = 0}} &= a_1(\mu_0) a_2(\mu_0) + a_1(\mu_0)^2 a_2(\mu_0) \\
&= -a_2(\mu_0) + a_2(\mu_0) \\
&= 0. \tag{4.12}
\end{aligned}
$$

From (4.11) and the Implicit Function Theorem, we thus conclude that near $\mu = \mu_0$ and $y = 0$, there exists a curve $\gamma: \ \mu = m(y)$ such that

$$
M(m(y), y) = 0, \tag{4.13}
$$

i.e., $\mu = m(y)$ represents period-2 points of the map f_μ. We have

$$
m(0) = \mu_0,
$$

and $m'(0)$, $m''(0)$ may be computed from (4.13):

$$
0 = \left[\frac{\partial M}{\partial \mu} m'(y) + \frac{\partial M}{\partial y} \right] \Big|_{\substack{y = 0 \\ \mu = m(0)}} = -2\alpha m'(0) + 0, \quad \text{(by (4.11) and (4.12))}, \tag{4.14}
$$

i.e., $m'(0) = 0$

$$
\left[\frac{\partial M}{\partial \mu} m''(y) + \frac{\partial^2 M}{\partial \mu^2} [m'(y)]^2 + 2 \frac{\partial M}{\partial \mu \partial y} m'(y) + \frac{\partial^2 M}{\partial y^2} \right] \Big|_{\substack{y = 0 \\ \mu = m(0)}} = 0,
$$

which implies

$$-2\alpha m''(0) + \frac{\partial^2 M(m(0), 0)}{\partial y^2} = -2\alpha m''(0) + 2[a_1(\mu_0)a_3(\mu_0)$$
$$+ 2a_1(\mu_0)a_2^2(\mu_0) + a_1^3(\mu_0)a_3(\mu_0)]$$
$$= -2\alpha m''(0) + 2 \cdot (-1)[2(a_3(\mu_0) + a_2^2(\mu_0))] = 0,$$

$$m''(0) = -\frac{2[a_3(\mu_0) + a_2^2(\mu_0)]}{\alpha} = -\frac{2}{\alpha}\left[\frac{1}{3!}\frac{\partial^3 g}{\partial y^3} + \left(\frac{1}{2!}\frac{\partial^2 g}{\partial y^2}\right)^2\right]\Bigg|_{\substack{y=0 \\ \mu=m(0)}}$$

$$= -\frac{2}{\alpha}\left[\frac{1}{3!}\frac{\partial^3 f}{\partial x^3} + \left(\frac{1}{2!}\frac{\partial^2 f}{\partial x^2}\right)^2\right]\Bigg|_{\substack{x=x_0 \\ \mu=\mu_0}}$$

$$= -\frac{2\beta}{\alpha} \neq 0.$$

Therefore, near $y = 0$, the function $\mu = m(y)$ has an expansion

$$\mu = m(0) + m'(0)y + \frac{m''(0)}{2!}y^2 + \mathcal{O}(|y|^3) \tag{4.15}$$

$$= \mu_0 - \frac{\beta}{\alpha}y^2 + \mathcal{O}(|y|^3). \quad \square \tag{4.16}$$

$\square$

Exercise 4.2   Verify that $\frac{\partial^3 g^2(\mu_0, 0)}{\partial y^3} = 3\frac{\partial^2 M(\mu_0, 0)}{\partial y^2} = -12\beta \neq 0$. $\square$

We now check the stability of the period-2 points by computing $\frac{\partial(g^2)}{\partial y}$ about $y = 0$ and $\mu = \mu_0$:

$$\frac{\partial(g^2)(\mu, y)}{\partial y} = \frac{\partial(g^2)(\mu_0, 0)}{\partial y} + \frac{\partial^2(g^2)(\mu_0, 0)}{\partial y^2}y + \frac{\partial^2(g^2)(\mu_0, 0)}{\partial\mu\partial y}(\mu - \mu_0)$$
$$+ \frac{1}{2}\frac{\partial^3(g^2)(\mu_0, 0)}{\partial y^3}y^2 + \cdots . \tag{4.17}$$

But

$$\frac{\partial(g^2)(\mu_0, 0)}{\partial y} = a_1^2(\mu_0) = (-1)^2 = 1, \tag{4.18}$$

$$\frac{\partial^2(g^2)(\mu_0, 0)}{\partial y^2} = a_1(\mu_0)a_2(\mu_0) + a_1^2(\mu_0)a(\mu_0)$$
$$= -a_2(\mu_0) + a_2(\mu_0) = 0, \tag{4.19}$$

$$\frac{\partial^2(g^2)(\mu_0, 0)}{\partial\mu\partial y} = 2a_1(\mu_0)a_1'(\mu_0) = -2\alpha, \quad \text{(by (4.10)),} \tag{4.20}$$

and

$$\frac{1}{2}\frac{\partial^3 (g^2)(\mu_0, 0)}{\partial y^3} = \frac{1}{2}(-12\beta) = -6\beta, \quad \text{(by Exercise 4.2)}. \tag{4.21}$$

Substituting (4.12) and (4.18)–(4.21) into (4.17), we obtain

$$\frac{\partial (g^2)(m(y), y)}{\partial y} = 1 + (-2\alpha) \cdot \left(-\frac{\beta}{\alpha}y^2\right) + (-6\beta)y^2 + \mathcal{O}(|y|^3)$$
$$= 1 - 4\beta y^2 + \mathcal{O}(|y|^3).$$

Therefore,

$$|1 - 4\beta y^2| = 1 - 4\beta y^2 < 1, \quad \text{if } \beta > 0, \quad \text{and so the period-2}$$
$$\text{orbit is } attracting;$$
$$|1 - 4\beta y^2| = 1 - 4\beta y^2 > 1, \quad \text{if } \beta < 0, \quad \text{and so the period-2}$$
$$\text{orbit is } repelling. \qquad \square$$

Exercise 4.3   Consider the iteration

$$x_{n+1} = \mu \sin(\pi x_n), \qquad 0 \le x_n \le 1, \quad 0 < \mu < 1.$$

(1) Use the Period Doubling Bifurcation Theorem to determine when (i.e., for what value of μ) the first period doubling happens.

(2) Determine the stability of the bifurcated period-2 solutions.

(3) Plot an orbit diagram to confirm your answers in (1) and (2). $\qquad \square$

4.2   SADDLE-NODE BIFURCATIONS

Next, we consider a different type of bifurcation.

Example 4.4   Let $f(\mu, x) = \mu e^x$, $x \in \mathbb{R}$. If $\mu > e^{-1}$, then the graph of $f(\mu, x)$ looks like Fig. 4.2(a), where there is no intersection between the curve $f(\mu, x)$ and the diagonal line $y = x$. However, when $\mu = e^{-1}$, $f(\mu, x)$ is tangent to the diagonal line at $x = 1$. See Fig. 4.2(b). For $\mu < e^{-1}$, the graph of $f(\mu, x)$ intersects $y = x$ at two points, both of which are then fixed points. Among these two fixed points, one is stable and the other is unstable. See Fig. 4.2(c). $\qquad \square$

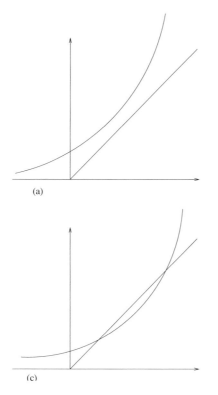

Figure 4.2: The graphs of $y = f(\mu, x) = \mu e^x$.
(a) $\mu > e^{-1}$;
(b) $\mu = e^{-1}$;
(c) $0 < \mu < e^{-1}$.

In Fig. 4.2(b), we see that the slope $\frac{\partial}{\partial x} f(\mu, x)$ is equal to 1 at the point of tangency where $x = 1$. There is a bifurcation of fixed points when $\mu = e^{-1}$ because $f(\mu, x)$ has changed behavior from having no fixed points in Fig. 4.2(a) to having two fixed points in Fig. 4.2(c). This bifurcation is now analyzed in the following theorem.

Theorem 4.5   (Saddle-Node or Tangent Bifurcation) Assume that $f(\cdot, \cdot) \colon \mathbb{R}^2 \to \mathbb{R}$ is C^2 satisfying the following conditions:

$$f(\mu_0, x_0) = x_0 \text{ and } \frac{\partial}{\partial x} f(\mu_0, x_0) = 1, \ \frac{\partial^2}{\partial x^2} f(\mu_0, x_0) \neq 0, \ \frac{\partial}{\partial \mu} f(x_0, \mu_0) \neq 0.$$

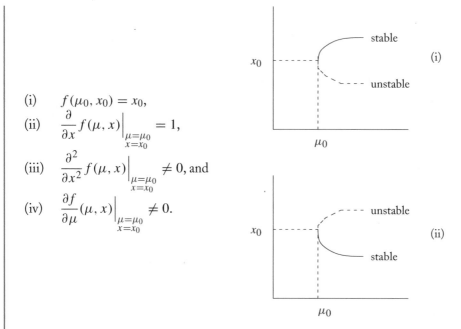

(i) $f(\mu_0, x_0) = x_0,$

(ii) $\dfrac{\partial}{\partial x} f(\mu, x)\Big|_{\substack{\mu=\mu_0 \\ x=x_0}} = 1,$

(iii) $\dfrac{\partial^2}{\partial x^2} f(\mu, x)\Big|_{\substack{\mu=\mu_0 \\ x=x_0}} \neq 0,$ and

(iv) $\dfrac{\partial f}{\partial \mu}(\mu, x)\Big|_{\substack{\mu=\mu_0 \\ x=x_0}} \neq 0.$

Figure 4.3: Saddle-node bifurcation.

Then there exists a curve $C: \mu = m(x)$ of fixed points defined in a neighborhood of x_0 such that $\mu_0 = m(x_0), m'(x_0) = 0,$ and

$$f(m(x), x) = x, \qquad m''(x_0) = -\frac{\partial^2 f/\partial x^2}{\partial f/\partial \mu}\Big|_{\substack{x=x_0 \\ \mu=\mu_0}} \neq 0. \qquad (4.22)$$

The fixed points on C are either (i) stable for $x > x_0$ and unstable for $x < x_0$, or (ii) stable for $x < x_0$ and unstable for $x < x_0$. (See Fig. 4.3.)

Proof. We follow Robinson [58, pp. 212–213]. To determine the fixed points, we define the function

$$G(\mu, x) = f(\mu, x) - x = 0.$$

Then $G(\mu_0, x_0) = 0.$ We have

$$\frac{\partial}{\partial x} G(\mu, x)\Big|_{\substack{\mu=\mu_0 \\ x=x_0}} = \left(\frac{\partial f}{\partial x} - 1\right)\Big|_{\substack{\mu=\mu_0 \\ x=x_0}} = 0,$$

so near $x = x_0$, x may not be solved in terms of y. However,

$$\frac{\partial}{\partial \mu} G(\mu, x)\bigg|_{\substack{\mu=\mu_0 \\ x=x_0}} = \frac{\partial f(\mu, x)}{\partial x}\bigg|_{\substack{\mu=\mu_0 \\ x=x_0}} \neq 0.$$

Thus, there exists a function $\mu = m(x)$ defined in a neighborhood of $x = x_0$ such that $\mu_0 = m(x_0)$ and

$$f(m(x), x) = x.$$

This function describes a curve C of fixed points near (μ_0, x_0). We have

$$0 = \frac{d}{dx} G(m(x), x) = \frac{\partial f}{\partial \mu} m'(x) + \frac{\partial f}{\partial x} - 1 = 0.$$

At $(\mu, x) = (\mu_0, x_0)$,

$$0 = \frac{\partial f(\mu_0, x_0)}{\partial \mu} m'(x_0) + \frac{\partial f(\mu_0, x_0)}{\partial x} - 1 = \frac{\partial f(\mu_0, x_0)}{\partial \mu} m'(x_0),$$

and so $m'(x_0) = 0$.

Also,

$$0 = \frac{d^2}{dx^2} G(m(x), x) = \frac{\partial^2 f}{\partial \mu^2} [m'(x)]^2 + 2\frac{\partial^2 f}{\partial \mu \partial x} m'(x) + \frac{\partial f}{\partial \mu} m''(x) + \frac{\partial^2 f}{\partial x^2} = 0. \qquad (4.23)$$

At $(\mu, x) = (\mu_0, x_0)$, using the fact that $m'(x_0) = 0$, $\frac{\partial f}{\partial \mu} \neq 0$, we obtain from (4.23) the second equation of (4.22).

Finally, we analyze stability of points on C near $(\mu, x) = (\mu_0, x_0)$. The stability is determined by whether

$$\left|\frac{\partial}{\partial x} f(\mu, x)\right|_{\substack{\mu=m(x) \\ x}} \quad \text{is less than 1 or greater than 1.}$$

We have

$$\frac{\partial}{\partial x} f(\mu, x)\bigg|_{\mu=m(x)} = \frac{\partial}{\partial x} f(\mu_0, x_0) + \frac{\partial^2 f(\mu_0, x_0)}{\partial x^2}(x - x_0)$$

$$+ \frac{\partial^2 f(\mu_0, x_0)}{\partial \mu \partial x}(m(x) - \mu_0)$$

$$+ \frac{1}{2!} \frac{\partial^3 f(\mu_0, x_0)}{\partial x^3}(x - x_0)^2$$

$$+ \frac{\partial^3 f(\mu_0, x_0)}{\partial \mu \partial x^2}(m(x) - \mu_0)(x - x_0)$$

$$+ \frac{1}{2!} \frac{\partial^3 f(\mu_0, x_0)}{\partial \mu^2 \partial x}(m(x) - \mu_0)^2 + \cdots$$

$$= 1 + \frac{\partial^2 f(\mu_0, x_0)}{\partial x^2}(x - x_0) + \mathcal{O}(|x - x_0|^2).$$

If $\frac{\partial^2 f(\mu_0, x_0)}{\partial x^2} > 0$, then

$$1 + \frac{\partial^2 f(\mu_0, x_0)}{\partial x^2}(x - x_0) \quad \begin{cases} > 1 & x > x_0, \\ & \text{if} \\ < 1 & x < x_0. \end{cases}$$

So the stability can be determined. A similar conclusion can be obtained if $\frac{\partial^2 f(\mu_0, x_0)}{\partial x^2} < 0$. $\square$

The Saddle-Node Bifurcation Theorem does not apply to the map $f(\mu, x) = \mu \sin(\pi x)$ we studied in Exercise 4.8. To analyze the bifurcation behavior of $f(\mu, x)$ near $x_0 = 0$ and $\mu_0 = 1/\pi$, we note that even though

$$\frac{\partial}{\partial x} f(\mu, x) \Big|_{\substack{x = x_0 = 0 \\ \mu = \mu_0 = 1/\pi}} = \mu_0 \pi \cos 0 = 1,$$

a different bifurcation must be happening, as

$$\frac{\partial}{\partial \mu} f(\mu, x) \Big|_{\substack{x = 0 \\ \mu = 1/\pi}} = \sin 0 = 0.$$

This will be investigated in the next section.

4.3   THE PITCHFORK BIFURCATION

The following theorem applies to the map $f(\mu, x) = \mu \sin(\pi x)$.

Theorem 4.6   (Pitchfork Bifurcation)   Let $f(\mu, x) = x g(\mu, x)$ and $f(\cdot, \cdot)$: $\mathbb{R}^2 \to \mathbb{R}$ is C^3. Then the curve C_1: $x \equiv 0$ represents a curve of fixed points on the (μ, x)-plane. Assume that at $(\mu, x) = (\mu_0, 0)$, we have

Then at $(\mu, x) = (\mu_0, 0)$, there is a new curve C_2: $\mu = m(x)$ of bifurcated fixed points such that

$$\mu_0 = m(0), \quad m'(0) = 0, \quad m''(0) \neq 0,$$
$$x = f(m(x), x).$$

The stability of the points on C_2 near $(\mu, x) = (\mu_0, 0)$ is attracting if $\partial^2 g(\mu_0, 0)/\partial x^2 < 0$ and repelling if $\partial^2 g(\mu_0, 0)/\partial x^2 > 0$.

Proof. Define the implicit relation

$$G(\mu, x) = g(\mu, x) - 1 = 0.$$

Note that if $x \neq 0$ satisfies

$$G(\mu, x) = g(\mu, x) - 1 = 0,$$

$$\frac{\partial}{\partial x} f(\mu, x)\bigg|_{\substack{\mu=\mu_0 \\ x=0}} = 1,$$

$$\frac{\partial}{\partial \mu} g(\mu, x)\bigg|_{\substack{\mu=\mu_0 \\ x=0}} \neq 0,$$

$$\frac{\partial^2}{\partial x^2} f(\mu, x)\bigg|_{\substack{\mu=\mu_0 \\ x=0}} = 0,$$

$$\frac{\partial^3}{\partial x^3} f(\mu, x)\bigg|_{\substack{\mu=\mu_0 \\ x=0}} \neq 0.$$

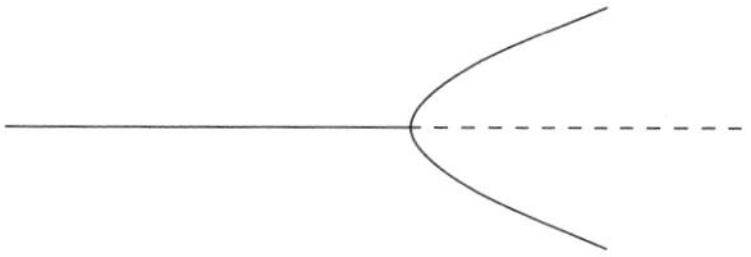

Figure 4.4: Pitchfork bifurcation.

then $xg(\mu, x) - x = 0 = f(\mu, x) - x$. Therefore, x is a fixed point of the map $f(\mu, x)$. Since

$$\frac{\partial}{\partial x} G(\mu, x)\bigg|_{\substack{\mu=\mu_0 \\ x=0}} = \frac{\partial}{\partial x} g(\mu, x)\bigg|_{\substack{\mu=\mu_0 \\ x=0}} = \frac{1}{2}\left[\frac{\partial^2}{\partial x^2} f(\mu, x)\bigg|_{\substack{\mu=\mu_0 \\ x=0}}\right] = 0,$$

we may not be able to solve x in terms of μ locally near $\mu = \mu_0$. However,

$$\frac{\partial}{\partial \mu} G(\mu, x)\bigg|_{\substack{x=0 \\ \mu=\mu_0}} = \frac{\partial g(\mu, x)}{\partial \mu}\bigg|_{\substack{\mu=\mu_0 \\ x=0}} \neq 0,$$

so we have a curve C_2: $\mu = m(x)$ of fixed points near $(\mu, x) = (\mu_0, 0)$, such that $m(0) = \mu_0$. Let us compute $m'(0)$ and $m''(0)$. We have

$$\frac{d}{dx} G(m(x), x) = 0 = \frac{\partial g}{\partial \mu} m'(x) + \frac{\partial g}{\partial x}.$$

Since $\partial g(\mu_0, 0)/\partial \mu \neq 0$,

$$m'(0) = -\frac{\partial g(\mu_0, 0)/\partial x}{\partial g(\mu_0, 0)/\partial \mu} = 0. \tag{4.24}$$

Differentiating again,

$$\frac{d^2}{dx^2} G(m(x), x) = 0 = \frac{\partial^2 g}{\partial \mu^2}[m'(x)]^2 + 2\frac{\partial^2 g}{\partial \mu \partial x} m'(x)$$
$$+ \frac{\partial g}{\partial \mu} m''(x) + \frac{\partial^2 g}{\partial x^2},$$

we obtain

$$m''(0) = -\frac{\partial^2 g(\mu_0, 0)/\partial x^2}{\partial g(\mu_0, 0)/\partial \mu}. \tag{4.25}$$

But
$$\frac{\partial^3}{\partial x^3} f(\mu, x) = \frac{\partial^3}{\partial x^3}[xg(\mu, x)] = 3\frac{\partial^2 g(\mu, x)}{\partial x^2} + x\frac{\partial^3 g(\mu, x)}{\partial x^3}$$

and at $x = 0$, $\mu = \mu_0$,
$$\frac{\partial^2 g(\mu_0, 0)}{\partial x^2} = \frac{1}{3}\frac{\partial^3}{\partial x^3} f(\mu_0, 0) \neq 0. \qquad (4.26)$$

From (4.25) and (4.26), we thus have
$$m''(0) \neq 0. \qquad (4.27)$$

Combining (4.24) and (4.27), we see that the curve C_2, locally, looks like a parabola near $(\mu_0, 0)$ on the (μ, x)-plane. C_2 opens to the left if $m''(0) < 0$, and to the right if $m''(0) > 0$.

Finally, let us analyze stability of the bifurcated fixed points on C_2 near $(\mu, x) = (\mu_0, 0)$. We have

$$\begin{aligned}
\left.\frac{\partial f(\mu, x)}{\partial x}\right|_{\mu=m(x)} &= \frac{\partial f(\mu_0, 0)}{\partial x} + \frac{\partial^2 f(\mu_0, 0)}{\partial x^2}(x - 0) + \frac{\partial^2 f(\mu_0, 0)}{\partial \mu \partial x}(m(x) - \mu_0) \\
&\quad + \frac{1}{2!}\frac{\partial^3 f(\mu_0, 0)}{\partial x^3}(x - 0)^2 + \frac{\partial^3 f(\mu_0, 0)}{\partial \mu \partial x^2}(x - 0)(m(x) - \mu_0) \\
&\quad + \frac{1}{2!}\frac{\partial^3 f(\mu_0, 0)}{\partial \mu^2 \partial x}(m(x) - \mu_0)^2 + \cdots \\
&= 1 + 0 \cdot x + \left[\frac{\partial g(\mu_0, 0)}{\partial \mu} \cdot m''(0)x^2 + \mathcal{O}(x^3)\right] \\
&\quad + \frac{1}{2}\left[3\frac{\partial^2 g(\mu_0, 0)}{\partial x^2}x^2\right] + \mathcal{O}(x^3). \qquad (4.28)
\end{aligned}$$

But, by (4.25),
$$\frac{\partial g(\mu_0, 0)}{\partial \mu}m''(0) = -\frac{\partial^2 g(\mu_0, 0)}{\partial x^2}, \qquad (4.29)$$

and by substituting (4.29) into (4.28), we obtain

$$\left.\frac{\partial f(\mu, x)}{\partial x}\right|_{\mu=m(x)} = 1 + \frac{1}{3}\frac{\partial^2 g(\mu_0, 0)}{\partial x^2}x^2 + \mathcal{O}(x^3).$$

Therefore, fixed points on C_2 near $(\mu, x) = (\mu_0, 0)$ are attracting if $\partial^2 g(\mu_0, 0)/\partial x^2 < 0$, and repelling if $\partial^2 g(\mu_0, 0)/\partial x^2 > 0$.   □

Example 4.7   For the map $f(\mu, x) = \mu \sin(\pi x)$, we have

$$f(\mu, x) = xg(\mu, x) \quad \text{where}$$
$$g(\mu, x) = \mu\left[\pi - \frac{\pi(\pi x)^2}{3!} + \frac{\pi(\pi x)^4}{5!} - \cdots + (-1)^n\frac{\pi(\pi x)^{2n}}{(2n+1)!} \pm \cdots\right].$$

At $(\mu, x) = \left(\frac{1}{\pi}, 0\right)$, we have

$$\frac{\partial f(\mu, x)}{\partial x} = 1, \quad \frac{\partial g(\mu, x)}{\partial \mu} = \pi \neq 0, \quad \frac{\partial^2 f(\mu, x)}{\partial x^2} = 0, \quad \frac{\partial^3 f(\mu, x)}{\partial x^3} = -\mu\pi^3 \neq 0,$$

$$\frac{\partial^2 g(\mu, x)}{\partial x^2} = -\frac{1}{3}\pi^2 < 0.$$

So the bifurcated fixed points are stable, as can be seen in Fig. 4.5. □

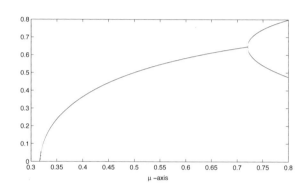

Figure 4.5: Bifurcation of the fixed points $x \equiv 0$ into a new curve C_2 of fixed points near $(\mu, x) = \left(\frac{1}{\pi}, 0\right)$, for the map $f(\mu, x) = \mu \sin(\pi, x), 0 \leq x \leq 1$.

Exercise 4.8   For the quadratic map

$$f(\mu, x) = \mu x(1 - x),$$

at $(\mu, x) = (1, 0)$, we have

$$\left.\frac{\partial}{\partial x} f(\mu, x)\right|_{\substack{\mu=1 \\ x=0}} = 1.$$

Analyze the bifurcation of fixed points near $(\mu, x) = (1, 0)$ by stating and proving a theorem similar to Theorem 4.6. □

4.4   HOPF BIFURCATION

To begin with, we offer as an example a 2-dimensional map $\boldsymbol{F} \colon \mathbb{R}^2 \to \mathbb{R}^2$ defined by

$$\boldsymbol{F}(x_1, x_2) = \begin{bmatrix} \cos\theta & -\sin\theta \\ \sin\theta & \cos\theta \end{bmatrix} \left\{ (1 + \alpha) \begin{bmatrix} x_1 \\ x_2 \end{bmatrix} + (x_1^2 + x_2^2) \begin{bmatrix} a & -b \\ b & a \end{bmatrix} \begin{bmatrix} x_1 \\ x_2 \end{bmatrix} \right\}, \tag{4.30}$$

where α is a parameter; $\theta = \theta(\alpha)$, $a = a(\alpha)$, $b = b(\alpha)$ are smooth functions of α, satisfying $0 < \theta(0) < \pi$, $a(0) \neq 0$, for $\alpha = 0$.

It is easy to check that the origin, $(x_1, x_2) = (0, 0)$, is a fixed point of $\boldsymbol{F}$ for all α. At (0,0), the Jacobian matrix of the map $\boldsymbol{F}$ is

$$A \equiv (1+\alpha) \begin{bmatrix} \cos\theta & -\sin\theta \\ \sin\theta & \cos\theta \end{bmatrix}.$$

The two eigenvalues of matrix A are $\mu_{1,2} \equiv (1+\alpha)e^{\pm i\theta}$. In particular, when $\alpha = 0$, we have $|\mu_{1,2}| = 1$. Thus, the origin is *not* a hyperbolic fixed point; cf. Def. 1.4. To facilitate the study of bifurcation of the system when α passes $\alpha = 0$, we rewrite $\boldsymbol{F}$ as a map of the complex plane:

for $z = x_1 + ix_2$, $x_1, x_2 \in \mathbb{R}$,
$$\boldsymbol{F}(z) = e^{i\theta}z(1 + \alpha + d|z|^2) = \mu z + cz|z|^2, \tag{4.31}$$
$$c = c(\alpha) \equiv e^{i\theta(\alpha)}d(\alpha), \quad d(\alpha) \equiv a(\alpha) + ib(\alpha), \quad \mu = \mu(\alpha) \equiv (1+\alpha)e^{i\theta(\alpha)}.$$

We look at the phase relation of (4.31): letting $z = \rho e^{i\phi}$ with $\rho = |z|$, we have

$$\boldsymbol{F}(z) = e^{i\theta}(\rho e^{i\phi})[1 + \alpha + (a + ib)\rho^2]$$
$$= \rho[(1 + \alpha + a\rho^2)^2 + b^2\rho^4]^{1/2}e^{i(\theta+\phi+\psi)},$$

where

$$\psi = \sin^{-1}\frac{b\rho^2}{[(1+\alpha+a\rho^2)^2 + b^2\rho^4]^{1/2}}.$$

Thus, in polar coordinates, system (4.31) becomes

$$\begin{bmatrix} \rho \\ \varphi \end{bmatrix} \longmapsto \boldsymbol{G}(\rho, \varphi) = \begin{bmatrix} \rho[1 + \alpha + a(\alpha)\rho^2] + \rho^4 R_\alpha(\rho) \\ \varphi + \theta(\alpha) + \rho^2 Q_\alpha(\rho) \end{bmatrix}, \tag{4.32}$$

where $R_\alpha(\cdot)$ and $Q_\alpha(\cdot)$ are smooth functions of (ρ, α). From (4.32), we know that the first component on the RHS of (4.32) is independent of φ. Thus, we have achieved *decoupling* between ρ and φ, making the subsequent discussions on bifurcation more intuitive. With regard to the ρ-variable, the transformation (4.31) actually constitutes a 1-dimensional dynamical system:

$$\widetilde{G}(p) \equiv \rho[1 + \alpha + a(\alpha)\rho^2] + \rho^4 R_\alpha(\rho).$$

For this dynamical system, $\rho = 0$ is a fixed point for any parameter value α. When $\alpha > 0$, the fixed point $\rho = 0$ is unstable. When $\alpha = 0$, the stability of $\rho = 0$ is determined by the sign of $a(0)$:

(i) if $a(0) < 0$, then $\rho = 0$ is (nonlinear) stable;

(ii) if $\alpha > 0$ and $a(0) < 0$, then in addition to the fixed point $\rho = 0$, there is another stable fixed point

$$\rho(\alpha) = \sqrt{-\frac{\alpha}{a(\alpha)}} + \mathcal{O}(\alpha). \tag{4.33}$$

With respect to the phase angle φ, the second component of the RHS of (4.32) shows that the action of the map is similar to a rotation by an angle $\theta(\alpha)$ (but it depends on both ρ and α).

Summarizing the above, we see that for the 2-dimensional dynamical system (4.30), assuming that $a(0) < 0$, then when the parameter α passes 0, we have the following bifurcation phenomena:

(1) When $\alpha < 0$, the origin (0,0) is a stable fixed point. The phase plot is a *stable focus*; see Fig. 4.6(a).

(2) When $\alpha > 0$, the origin (0,0) is an *unstable focus* in a small neighborhood of the origin. There is a closed curve C with approximate radius $\rho(\alpha)$ (see (4.33)), which is a stable curve such that the trajectories of all points starting from either within C or outside C will be attracted to C; see Fig 4.6(c).

(3) When $\alpha = 0$, the origin (0,0) is nonlinear stable; cf. Fig. 4.6(b).

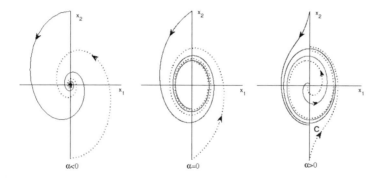

Figure 4.6: Hopf bifurcation.

The above bifurcation phenomena are called the *Hopf bifurcation* (or the *Neimark–Sacker bifurcation*). *Its special feature is the occurrence of the closed curve C, which is invariant* under the map $\boldsymbol{F}$. Similarly, one may consider the case $a(0) > 0$. For this case, there is an unstable closed curve C when $\alpha > 0$, Hopf bifurcation happens when α crosses 0, and the curve C disappears.

Next, we consider the following two-dimensional map:

$$\boldsymbol{F}(x_1, x_2) = \begin{bmatrix} \cos\theta & -\sin\theta \\ \sin\theta & \cos\theta \end{bmatrix} \left\{ (1+\alpha) \begin{bmatrix} x_1 \\ x_2 \end{bmatrix} + (x_1^2 + x_2^2) \begin{bmatrix} a & -b \\ b & a \end{bmatrix} \begin{bmatrix} x_1 \\ x_2 \end{bmatrix} \right\}$$
$$+ \mathcal{O}(|x|^4). \tag{4.34}$$

Similarly, to the conversion from (4.30) to (4.31), one can assert that (4.34) can be converted to the form

$$z \longmapsto F(z) = \mu z + cz|z|^2 + \mathcal{O}(|z|^4). \tag{4.35}$$

In comparison, system (4.34) contains some higher order terms than those of system (4.30). Even though (4.34) is not locally topologically conjugate to (4.30), and the higher order terms in (4.34) do affect bifurcation phenomena, some special characters of Hopf bifurcation are preserved.

Lemma 4.9   The higher order term $\mathcal{O}(|x|^4)$ in (4.34) does not affect the occurrence (or disappearance) of the invariant closed close C and its stability. The local stability of the origin $(0,0)$ and the bifurcation patterns remain the same.

Proof. The justification is lengthy. We refer to Kuznetsov [46, pp. 131–136], for example. □

Now, we consider a general planar map

$$f \colon \mathbb{R}^2 \to \mathbb{R}^2, \quad x \longmapsto f(x, \alpha), \quad x = (x_1, x_2) \in \mathbb{R}^2, \quad \alpha \in \mathbb{R}. \tag{4.36}$$

We will prove that any planar map with the Hopf bifurcation property can be transformed into the form (4.34).

Assume that f is smooth, and at $\alpha = 0$, $f(x, \alpha)$ has a fixed point $x = 0$, i.e., $f(0, 0) = 0$. The eigenvalues of the Jacobian matrix at $(x, \alpha) = (0, 0)$ are $\mu_{1,2} = e^{\pm i\theta_0}$, for some $\theta_0 \colon 0 < \theta_0 < \pi$. By the Implicit Function Theorem, for sufficiently small $|\alpha|$, f has a one-parameter family of unique fixed points $x(\alpha)$, and the map f is invertible. By using translation, the fixed points $x(\alpha)$ can be relocated to the origin 0. Thus, without loss of generality, we assume that for $|\alpha|$ small, $x = 0$ is a fixed point of the system. Thus, the map can be written as

$$x \longmapsto A(\alpha)x + F(x, \alpha), \tag{4.37}$$

where $A(\alpha)$ is a 2×2 matrix depending on α, and $F(x, \alpha)$ is a vector valued function with components F_1 and F_2 such that their leading terms begin quadratically with respect to x_1 and x_2 in the Taylor expansions of F_1 and F_2, and $F(0, \alpha) = 0$ for sufficiently small $|\alpha|$. The Jacobian matrix $A(\alpha)$ has eigenvalues

$$\mu_{1,2}(\alpha) = r(\alpha)e^{\pm i\varphi(\alpha)}, \text{ with } r(0) = 1, \varphi(0) = \theta_0. \tag{4.38}$$

Set $\beta(\alpha) = r(\alpha) - 1$. Then $\beta(\alpha)$ is smooth and $\beta(0) = 0$. Assume that $\beta'(0) \neq 0$, then (locally) we can use β in lieu of α for the parametrization. Thus, we have

$$\mu_1(\beta) = \mu(\beta) \equiv (1 + \beta)e^{i\theta(\beta)}, \quad \mu_2(\beta) = \bar{\mu}(\beta), \tag{4.39}$$

where $\theta(\beta)$ is a smooth function of β and $\theta(0) = \theta_0$.

Lemma 4.10   Under the assumptions of (4.38) and (4.39) for $|\alpha|$ small, the map (4.37) can be rewritten as

$$z \longmapsto \mu(\beta)z + g(z, \bar{z}, \beta), \quad z \in \mathbb{C}, \quad \beta \in \mathbb{R}, \tag{4.40}$$

where g is a smooth function with a (local) Taylor's expansion

$$g(z, \bar{z}, \beta) = \sum_{k+\ell \geq 2} \frac{1}{k!\ell!} g_{k\ell}(\beta) z^k \bar{z}^\ell, \qquad k, \ell = 0, 1, 2, \dots .$$

Proof. Let $\boldsymbol{q}(\beta)$ be the eigenvector corresponding to eigenvalue $\mu(\beta)$:

$$A(\beta)\boldsymbol{q}(\beta) = \mu(\beta)\boldsymbol{q}(\beta). \tag{4.41}$$

Then $A^T(\beta)$, the transpose of $A(\beta)$, has an eigenvector $\boldsymbol{p}(\beta)$ corresponding to eigenvalue $\overline{\mu}(\beta)$:

$$A^T(\beta)\boldsymbol{p}(\beta) = \overline{\mu}(\beta)\boldsymbol{p}(\beta).$$

First, we prove that

$$\langle \boldsymbol{p}, \bar{\boldsymbol{q}} \rangle \equiv \langle \boldsymbol{p}, \bar{\boldsymbol{q}} \rangle_{\mathbb{C}^2} = 0,$$

where $\langle p, q \rangle$ is defined as

$$\langle p, q \rangle = \bar{p}_1 q_1 + \bar{p}_2 q_2 \quad \text{for} \quad p = (p_1, p_2)^T, \quad q = (q_1, q_2)^T.$$

In fact, since $A\boldsymbol{q} = \mu\boldsymbol{q}$ and A is a real matrix, we have $A\bar{\boldsymbol{q}} = \bar{\mu}\bar{\boldsymbol{q}}$. Thus,

$$\begin{aligned}
\langle \boldsymbol{p}, \bar{\boldsymbol{q}} \rangle &= \left\langle \boldsymbol{p}, \frac{1}{\bar{\mu}} A\bar{\boldsymbol{q}} \right\rangle \\
&= \frac{1}{\bar{\mu}} \langle \boldsymbol{p}, A\bar{\boldsymbol{q}} \rangle = \frac{1}{\bar{\mu}} \langle A^T \boldsymbol{p}, \bar{\boldsymbol{q}} \rangle \\
&= \frac{1}{\bar{\mu}} \langle \bar{\mu}\boldsymbol{p}, \bar{\boldsymbol{q}} \rangle = \frac{\mu}{\bar{\mu}} \langle \boldsymbol{p}, \bar{\boldsymbol{q}} \rangle.
\end{aligned}$$

Therefore,

$$\left(1 - \frac{\mu}{\bar{\mu}} \right) \langle \boldsymbol{p}, \bar{\boldsymbol{q}} \rangle = 0.$$

But $\mu \neq \bar{\mu}$ because for sufficiently small $|\beta|$, we have $0 < \theta(\beta) < \pi$. So we have

$$\langle \boldsymbol{p}, \bar{\boldsymbol{q}} \rangle = 0.$$

Second, since A is a real matrix and the imaginary part of μ is nonzero, the imaginary part of $\boldsymbol{q}$ is also nonzero. By the above equality, we have

$$\langle \boldsymbol{p}, \boldsymbol{q} \rangle \neq 0.$$

By normalization, we can assume that

$$\langle \boldsymbol{p}, \boldsymbol{q} \rangle = 1.$$

For any sufficiently small $|\beta|$, any $\boldsymbol{x} \in \mathbb{R}^2$, there exists a unique $z \in \mathbb{C}$ such that

$$\boldsymbol{x} = z\boldsymbol{q}(\beta) + \overline{z}\,\overline{\boldsymbol{q}}(\beta). \tag{4.42}$$

(Due to the fact that $\langle \boldsymbol{p}(\beta), \overline{\boldsymbol{q}}(\beta) \rangle_{\mathbb{C}^2} = 0$, we can simply choose $z = \langle \boldsymbol{p}(\beta), \boldsymbol{x} \rangle_{\mathbb{C}^2}$.)
From (4.37) and (4.41) through (4.42), for the complex variable z, we have

$$z \longmapsto \mu(\beta)z + \langle \boldsymbol{p}(\beta), F(z\boldsymbol{q}(\beta) + \overline{z}\,\overline{\boldsymbol{q}}(\beta), \beta) \rangle_{\mathbb{C}^2}. \tag{4.43}$$

Denote the very last term in (4.43) as $g(z, \overline{z}, \beta)$. Then we obtain

$$g(z, \overline{z}, \beta) = \sum_{k+\ell \geq 2} \frac{1}{k!\ell!} g_{k\ell}(\beta) z^k \overline{z}^\ell,$$

where

$$g_{k\ell}(\beta) = \frac{\partial^{k+\ell}}{\partial z^k \partial \overline{z}^\ell} \langle \boldsymbol{p}(\beta), F(z\boldsymbol{q}(\beta) + \overline{z}\,\overline{\boldsymbol{q}}(\beta), \beta) \rangle|_{z=0},$$

for $k + \ell \geq 2, k, \ell = 0, 1, 2, \ldots$. Hence, (4.40) is obtained. $\qquad \square$

The following three lemmas show that under proper conditions, we can convert the map from the form (4.40) to the standard form (4.36).

Lemma 4.11   Assume that $e^{i\theta_0} \neq 1, e^{3i\theta_0} \neq 1$. Consider the map

$$z \longmapsto \mu z + \frac{g_{20}}{2} z^2 + g_{11} z\overline{z} + \frac{g_{02}}{2} \overline{z}^2 + \mathcal{O}(|z|^3), \qquad z \in \mathbb{C}, \tag{4.44}$$

where $\mu = \mu(\beta) = (1 + \beta)e^{i\theta(\beta)}$, $g_{ij} = g_{ij}(\beta)$. Then for $|\beta|$ sufficiently small, there exists a (locally) invertible transformation

$$z = w + \frac{h_{20}}{2} w^2 + h_{11} w\overline{w} + \frac{h_{02}}{2} \overline{w}^2, \tag{4.45}$$

such that (4.44) is transformed to

$$w \longmapsto \mu w + \mathcal{O}(|w|^3), \qquad w \in \mathbb{C},$$

i.e., the quadratic terms $\mathcal{O}(|z|^2)$ in (4.44) are eliminated.

Proof. It is easy to check that (4.45) is invertible near the origin, as

$$w = z - \left(\frac{h_{20}}{2} z^2 + h_{11} z\overline{z} + \frac{h_{02}}{2} \overline{z}^2 \right) + \mathcal{O}(|z|^3).$$

With respect to the new complex variable w, (4.45) becomes

$$\tilde{w} \equiv \mu w + \frac{1}{2}[g_{20} + (\mu - \mu^2)h_{20}]w^2 = [g_{11} + (\mu - |\mu|^2)h_{11}]w\overline{w}$$
$$+ \frac{1}{2}[g_{02} + (\mu - \overline{\mu})^2 h_{02}]\overline{w}^2 + \mathcal{O}(|w|^3). \tag{4.46}$$

As

$$\mu^2(0) - \mu(0) = e^{i\theta_0}(e^{i\theta_0} - 1) \neq 0,$$
$$|\mu(0)|^2 - \mu(0) = 1 - e^{i\theta_0} \neq 0$$
$$\overline{\mu}(0)^2 - \mu(0) = e^{i\theta_0}(e^{-i3\theta_0} - 1) \neq 0,$$

for $|\beta|$ sufficiently small, we thus can let

$$h_{20} = \frac{g_{20}}{\mu^2 - \mu}, \quad h_{11} = \frac{g_{11}}{|\mu|^2 - \mu}, \quad h_{02} = \frac{g_{02}}{\overline{\mu}^2 - \mu}.$$

Hence, all the quadratic terms in (4.47) disappear. The proof is complete. $\qquad\square$

Remark 4.12

(i) Denote $\mu_0 = \mu(0)$. Then the conditions $e^{i\theta_0} \neq 1$ and $e^{3i\theta_0} \neq 1$ in Lemma 4.11 mean that $\mu_0 \neq 1, \mu_0^3 \neq 1$. The condition $\mu_0 \neq 1$ is automatically satisfied as $\mu_0 = e^{i\theta_0}$ and $0 < \theta_0 < \pi$.

(ii) From the transformation (4.45), we see that in the neighborhood of the origin, it is nearly an identity transformation.

(iii) The transformation (4.45) generally alters the coefficients of the cubic terms. $\qquad\square$

Lemma 4.13   Assume that $e^{2i\theta_0} \neq 1, e^{4i\theta_0} \neq 1$. Consider the map

$$z \longmapsto \mu z + \left[\frac{g_{30}}{6}z^3 + \frac{g_{21}}{2}z^2\overline{z} + \frac{g_{12}}{2}z\overline{z}^2 + \frac{g_{03}}{6}\overline{z}^3\right] + \mathcal{O}(|z|^4), \tag{4.47}$$

where $\mu = \mu(\beta) = (1 + \beta)e^{i\theta(\beta)}$, $g_{ij} = g_{ij}(\beta)$. For $|\beta|$ small, the following transformation

$$z = w + \left[\frac{h_{30}}{6}w^3 + \frac{h_{21}}{2}w^2\overline{w} + \frac{h_{12}}{2}w\overline{w}^2 + \frac{h_{03}}{6}\overline{w}^3\right] \tag{4.48}$$

converts (4.47) to

$$w \longmapsto \mu w + \frac{g_{21}}{2}w^2\overline{w} + \mathcal{O}|w|^4), \tag{4.49}$$

i.e., only one cubic term is retained in (4.49).

Proof. The map is locally invertible near the origin:

$$w = z - \left[\frac{h_{30}}{6}z^3 + \frac{h_{21}}{2}z^2\bar{z} + \frac{h_{12}}{2}z\bar{z}^2 + \frac{h_{03}}{6}\bar{z}^3\right] + \mathcal{O}(|z|^4).$$

Substituting (4.48) into (4.47), we obtain

$$\tilde{w} \equiv \mu w + \left\{\frac{1}{6}[g_{30} + (\mu - \mu^3)h_{30}]w^3 + \frac{1}{2}[g_{21} + (\mu - \mu|\mu|^2)h_{21}]w^2\bar{w}\right.$$
$$\left. +\frac{1}{2}[g_{12} + (\mu - \bar{\mu}|\mu|^2)h_{12}]w\bar{w}^2 + \frac{1}{6}[g_{03} + (\mu - \bar{\mu}^3)h_{03}]\bar{w}^3\right\} + \mathcal{O}(|w|^4).$$

If we set

$$h_{30} = \frac{g_{30}}{\mu^3 - \mu}, \quad h_{12} = \frac{g_{12}}{\bar{\mu}|\mu|^2 - \mu}, \quad h_{03} = \frac{g_{03}}{\bar{\mu}^3 - \mu},$$

which is viable as the denominators are nonzero by assumption, as well as $h_{21} = 0$, then we obtain (4.49).   □

The terms $\frac{g_{21}}{2}w^2\bar{w}$ in (4.49) is called the *resonance* terms. Note that its coefficient, $g_{21}/2$, is the same as the corresponding term in (4.47).

Lemma 4.14   (Normal form of Hopf bifurcation) Assume that $e^{ik\theta_0} \neq 1$ for $k = 1, 2, 3, 4$. Consider the map

$$z \longmapsto \mu z + \left(\frac{g_{20}}{2}z^2 + g_{11}z\bar{z} + \frac{g_{02}}{2}\bar{z}^2\right) + \left(\frac{g_{30}}{6}z^3 + \frac{g_{21}}{2}z^2\bar{z} + \frac{g_{12}}{2}z\bar{z}^2 + \frac{g_{03}}{2}\bar{z}^3\right)$$
$$+ \mathcal{O}(|z|^4), \tag{4.50}$$

where $\mu = \mu(\beta) = (1 + \beta)e^{i\theta(\beta)}, g_{ij} = g_{ij}(\beta), \theta_0 = \theta(\beta)|_{\beta=0}$. Then there exists a locally invertible transformation near the origin:

$$z = w + \left(\frac{h_{20}}{2}w^2 + h_{11}w\bar{w} + \frac{h_{02}}{2}\bar{w}^2\right) + \left(\frac{h_{30}}{6}w^3 + \frac{h_{12}}{2}w\bar{w}^2 + \frac{h_{03}}{6}\bar{w}^3\right)$$

such that for $|\beta|$ sufficiently small, (4.50) is transformed to

$$w \longmapsto \mu w + c_1 w^2\bar{w} + \mathcal{O}(|w|^4),$$

where

$$c_1 = c_1(\beta) = \frac{g_{20}g_{11}(\bar{\mu} - 3 + 2\mu)}{2(\mu^2 - \mu)(\bar{\mu} - 1)} + \frac{|g_{11}|^2}{1 - \bar{\mu}} + \frac{|g_{02}|^2}{2(\mu^2 - \bar{\mu})} + \frac{g_{21}}{2}. \tag{4.51}$$

Proof. This follows as a corollary to Lemmas 4.11 and 4.13. The value c_1 in (4.51) can be obtained by straightforward calculations.   □

We can now summarize all of the preceding discussions in this section. As the map (4.34) is (4.35) in essence, from Lemma 4.9, we obtain the Hopf bifurcation theorem for general planar maps as follows.

Theorem 4.15   (Hopf–(Neimark–Sacker) Bifurcation Theorem) For the 1-parameter family of planar maps

$$\boldsymbol{x} \longmapsto f(\boldsymbol{x}, \alpha),$$

assume that

(i) When $\alpha = 0$, the system has a fixed point $\boldsymbol{x}_0 = \boldsymbol{0}$, and the Jacobian matrix has eigenvalues

$$\mu_{1,2} = e^{\pm i\theta_0}, \qquad 0 < \theta_0 < \pi.$$

(ii) $r'(0) \neq 0$, where $r(\alpha)$ is defined through (4.38).

(iii) $e^{ik\theta_0} \neq 1$, for $k = 1, 2, 3, 4$.

(iv) $a(0) \neq 1$, where $a(0) = \mathrm{Re}(e^{-i\theta_0}c_1(0))$, with c_1 as given in (4.51).

Then when α passes through 0, the system has a closed invariant curve C bifurcating from the fixed point $\boldsymbol{x}_0 = \boldsymbol{0}$.   □

In applications, we often want to obtain the actual value of $a(0)$, which, from (4.51):

$$c_1(0) = \frac{g_{20}(0)g_{11}(0)(1 - 2\mu_0)}{2(\mu_0^2 - \mu_0)} + \frac{|g_{11}(0)|^2}{1 - \overline{\mu_0}} + \frac{|g_{02}(0)|^2}{2(\mu^2 - \overline{\mu_0})} + \frac{g_{21}(0)}{2},$$

is

$$a(0) = \mathrm{Re}\left(\frac{e^{-i\theta_0}g_{21}(0)}{2}\right) - \mathrm{Re}\left[\frac{(1 - 2e^{i\theta_0})e^{-2i\theta_0}}{2(1 - e^{i\theta_0})}g_{20}(0)g_{11}(0)\right] - \frac{1}{2}|g_{11}(0)|^2$$
$$- \frac{1}{4}|g_{02}(0)|^2.$$

NOTES FOR CHAPTER 4

The word bifurcation or *Abzweigung* (German) seems to have been first introduced by the celebrated German mathematician Carl Jacobi (1804-1851) [44] in 1834 in his study of the bifurcation of the McLaurin spheroidal figures of equilibrium of self-gravitating rotating bodies (Abraham and Shaw [1, p. 19], Iooss and Joseph [43, p. 11]). Poincare introduced the French word *bifurcation* in [57] in 1885. The bifurcation theorems studied in this chapter are of the *local character*, namely, local bifurcations, which analyze changes in the local stability properties of equilibrium points, periodic points or orbits or other invariant sets as system parameters cross through certain critical thresholds. The analysis of change of stability and bifurcation is almost always technical. No more so than the case of maps when the governing system consists of ordinary differential equations or even partial differential equations.

A partial list of reference sources for the study of bifurcations of maps, ordinary and partial differential equations are Hale and Kocak [31], Iooss and Joseph [43], Guckenheimer and Holmes [30], Robinson [58], and Wiggins [68, 69].

CHAPTER 5

Homoclinicity. Lyapunoff Exponents

5.1 HOMOCLINIC ORBITS

There is a very important *geometric* concept, called *homoclinic orbits*, that leads to chaos.

Let p be a fixed point of a C^1-map f:

$$f(p) = p.$$

Assume that p is repelling so that $|f'(p)| > 1$. Since p is repelling, there is a neighborhood $N(p)$ of p such that

$$|f(x) - f(p)| = |f(x) - p| > |x - p|, \qquad \forall x \in N(p). \tag{5.1}$$

We denote $W^u_{\text{loc}}(p)$ the largest open neighborhood of p such that (5.1) is satisfied. $W^u_{\text{loc}}(p)$ is called the *local unstable set of p*.

Definition 5.1   Let p be a repelling fixed point of a continuous map f, and let $W^u_{\text{loc}}(p)$ be the local unstable set of p. Let $x_0 \in W^u_{\text{loc}}(p)$. We say that x_0 is *homoclinic* to p if there exists a positive integer n such that

$$f^n(x_0) = p.$$

We say that x_0 is *heteroclinic* to p if there exists another different periodic point q such that

$$f^m(x_0) = q. \qquad \square$$

See some illustrations in Fig. 5.1(a) and (b).

Definition 5.2   A homoclinic orbit is said to be *nondegenerate* if $f'(x) \neq 0$ for all x on the orbit. Otherwise, it is said to be *degenerate*. (See Fig. 5.2.) $\qquad \square$

A nondegenerate homoclinic orbit will lead to chaos, as the following theorem shows.

Theorem 5.3   Let I be a bounded closed interval and $f : I \to I$ is C^1. Assume that p is a repelling fixed point of f, and p has a nondegenerate homoclinic orbit. Then

$$V_I(f^n) \geq K e^{\alpha n} \to \infty \quad \text{as} \quad n \to \infty, \tag{5.2}$$

for some K and $\alpha > 0$.

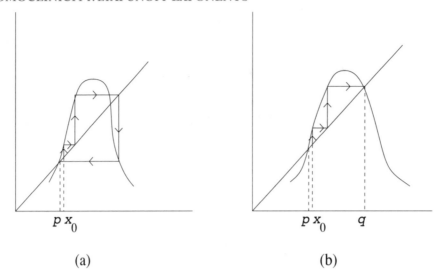

(a)   (b)

Figure 5.1: (a) p is a fixed point and $x_0 \in W^u_{\text{loc}}(p)$. But $f^4(x_0) = p$. So x_0 is homoclinic to p. (b) p and q are two different fixed points and $x_0 \in W^u_{\text{loc}}(p)$. But $f^3(x_0) = q$. So x_0 is heteroclinic to p.

Proof. Let $x_0 \in W^u_{\text{loc}}(p)$ such that $f^n(x_0) = p$, for some positive integer n. Since x_0 is nondegenerate,

$$[f^n(x)]'_{x=x_0} = f'(f^{n-1}(x_0)) \cdot f'(f^{n-2}(x_0)) \cdot \ \cdots \ \cdot f'(f(x_0)) \cdot f'(x_0) \neq 0. \qquad (5.3)$$

But assumption, $|f'(p)| > 1$. We can choose an open set $W \subseteq W^u_{\text{loc}}(p)$ such that

$$|f'(x)| \geq d > 1, \quad \forall x \in W, \quad \text{and} \quad p \in W. \qquad (5.4)$$

Now, choose an open interval $V \ni x_0$ such that $p \notin V$. Then by (5.3), if we choose V sufficiently small, we have

$$(f^n)'(x) \neq 0 \qquad \forall x \in V. \qquad (5.5)$$

This implies that $f, f^2, \ldots, f^n$ are 1-1 on V and, therefore,

$$V, f(V), f^2(V), \ldots, f^n(V)$$

are all open intervals, with $x_0 \in V$ and $p \in f^n(V)$. Furthermore, by choosing V sufficiently small, we may assume that

$$f^j(V) \cap f^k(V) = \emptyset \quad \text{for} \quad j \neq k, \quad j, k = 0, 1, 2 \ldots, n. \qquad (5.6)$$

If V is sufficiently small, then

$$f^n(V) \subseteq W. \qquad (5.7)$$

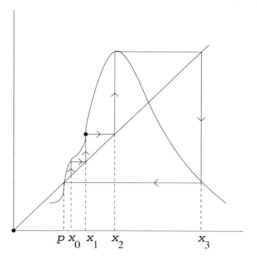

Figure 5.2: The orbit of x_0 is a degenerate homoclinic orbit because $f'(x_2) = 0$.

By the fact that the orbit of x_0 is nondegenerate, we have

$$|(f^j)'(x)| \geq \varepsilon \quad \text{for all} \quad j = 1, 2, \ldots, n, \quad \forall x \in V. \tag{5.8}$$

From (5.4), (5.6) and (5.7), we now have

$$\begin{aligned} |(f^{n+k})'(x)| &= |f'(f^{n+k-1}(x))| \cdot |f'(f^{n+k-2}(x))| \cdots |f'(f(x))||f'(x)| \\ &\geq d^k \varepsilon^n \end{aligned} \tag{5.9}$$
$$\forall x \in V, \quad \text{provided that } f^{n+j}(x) \in W \text{ for } j = 1, 2, \ldots, k.$$

Since n in (5.9) is fixed, by choosing k sufficiently large, we have some k such that

$$d^k \varepsilon^n \geq M \tag{5.10}$$

for any given $M > 1$. This implies that

$$f^{n+k}(V) \supseteq V$$

if k is chosen sufficiently large.

We now choose an open interval $V_1 \ni p$, $V_1 \subseteq W$, $V_1 \cap V = \emptyset$ and, by (5.10), we can choose k sufficiently large such that

$$f^{n+k}(V_1) \supseteq V.$$

Then by choosing V_1 sufficiently small, we have

$$f^{n+k}(V) \supseteq V_1 \cup V.$$

Therefore, we obtain

$$\overline{V}_1 \xrightarrow{f^{n+k}} \overline{V} \xrightarrow{f^{n+k}} \overline{V}_1 \cup \overline{V}. \tag{5.11}$$

This gives the growth of total variations

$$V_{\overline{V}_1}((f^{n+k})^j) \geq K' e^{\alpha' j} \to \infty \text{ as } j \to \infty, \text{ for some } K', \alpha' > 0,$$

where $V_{\overline{V}_1}$ denotes the total variation over the set $\overline{V}_1$. Using the above, we can further show that

$$V_{\overline{V}_1}((f^{n+k})^j \circ f^\ell) \geq K e^{\alpha j}, \text{ for } \ell = 1, 2, \ldots, n+k-1,$$

for some $K, \alpha > 0$.

Therefore, (5.1) has been proven.   □

Usually, the covering-interval sequence is much "stronger" than what (5.11) indicates. See Fig. 5.3.

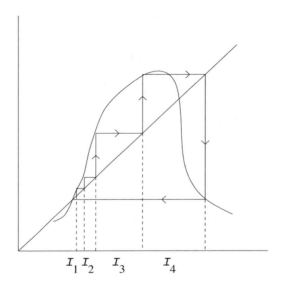

Figure 5.3: There is a homoclinic orbit showing the interval covering relation $I_1 \longrightarrow I_2 \longrightarrow I_3 \longrightarrow I_4 \longrightarrow I_1 \cup I_2 \cup I_3$.

Theorem 5.3 tells us that chaos occur when there is a nondegenerate homoclinic orbit. What happens if, instead, a homoclinic orbit is *degenerate*? This happens, e.g., for the quadratic map $f_\mu(x) = \mu x(1-x)$ when $\mu = 4$, and $x = 1/2$ lies on a degenerate homoclinic orbit.

In this case, the map has rather complex bifurcation behavior. For example, for the quadratic map f_μ mentioned above, near μ (actually, for $\mu > 4$) there are μ-values where there are infinitely many distinct homoclinic orbits. The maps f_μ also have sadddle-node or period-doubling bifurcations (Devaney [20]), which means that these bifurcations are accumulation points of simple bifurcations. This phenomenon is called *homoclinic bifurcation*; see also Afraimovich and Hsu ([2, pp. 195–208]).

5.2   LYAPUNOFF EXPONENTS

Let $f: I \to I$ be C^1 everywhere except at finitely many points on I. The *Lyapunoff exponent* of f at $x_0 \in \mathbb{R}$ is defined by

$$\lambda(x_0) = \limsup_{n \to \infty} \frac{1}{n}[\ln|(f^n)'(x_0)|]$$

if it exists. Because

$$(f^n)'(x_0) = f'(x_{n-1})f'(x_{n-2}) \cdots f'(x_1)f'(x_0), \quad \text{where} \quad x_j = f^j(x_0),$$

we have

$$\lambda(x_0) = \limsup_{n \to \infty} \frac{1}{n}\left[\sum_{j=0}^{n-1} \ln|f'(f^j(x_0))|\right].$$

Notation. Given x_0, denote

$$\mathcal{O}^+(x_0) = \text{the forward orbit of } x_0$$
$$= \{f^j(x_0) \mid j = 0, 1, 2, \ldots\};$$
$$\mathcal{O}^-(x_0) = \text{a backward orbit of } x_0$$
$$= \{f^{-j}(x_0) \mid j = 0, 1, 2, \ldots\}. \qquad \square$$

Example 5.4   The roof (or tent) function; see Fig. 5.4.

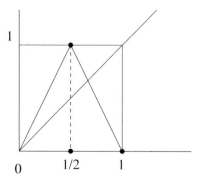

Figure 5.4: The roof function defined by $T(x) = \{2x, 0 \leq x \leq 1/2, 2(1-x), 1/2 < x \leq 1.$

At $x = 1/2$, $T(x)$ is not differentiable. So remove the set $\mathcal{O}^-(1/2)$. Choose any $y_0 \notin \mathcal{O}^-(1/2)$. Then $T(x)$ is differentiable at any $T^j(y_0)$. We have the Lyapunoff exponent

$$\lambda(y_0) = \limsup_{n \to \infty} \frac{1}{n} \left[\sum_{j=0}^{n-1} \ln |T'(T^j(y_0))| \right]$$

$$= \limsup_{n \to \infty} \frac{1}{n} \underbrace{[\ln |2| + \ln |2| + \cdots + \ln |2|]}_{n \text{ terms}} = \ln 2.$$

This suggests that when chaos occurs, $\lambda(y_0) > 0$.   □

Example 5.5   The quadratic map

$$f_\mu(x) = \mu x(1 - x), \qquad 3 < \mu < 3 + \delta_0,$$

where $3 + \delta_0$ is the μ-value where the second period-doubling bifurcation happens. See Fig. 5.5.

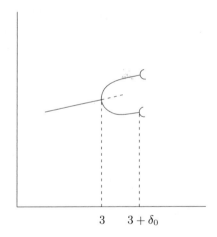

Figure 5.5: Second period-doubling bifurcation of the period-two points.

At $x = 1/2$, $f_\mu' \left(\frac{1}{2} \right) = 0$ and $\ln \left| f_\mu' \left(\frac{1}{2} \right) \right| = -\infty$. So we need to exclude $\mathcal{O}^-(1/2)$. If $y_0 \notin \mathcal{O}^-(1/2)$, then $f'(f_\mu^j(y_0)) \neq 0$ for any $j = 0, 1, 2, \ldots$. Note that if $y_0 \notin \{0\} \cup \mathcal{O}^-(1)$, then y_0 will be attracted to the period-2 orbit $\{z_0, f_\mu(z_0)\}$, which is globally attracting, i.e., either

$$\lim_{n \to \infty} f_\mu^{2n}(y_0) = z_0 \qquad \left(\text{and } \lim_{n \to \infty} f_\mu^{2n+1}(y_0) = f(z_0) \right) \qquad (5.12)$$

or

$$\lim_{n \to \infty} f_\mu^{2n+1}(y_0) = z_0 \qquad \left(\text{and } \lim_{n \to \infty} f_\mu^{2n}(y_0) = f(z_0) \right). \qquad (5.13)$$

Then we have (check!)

$$\lambda(y_0) = \max \left\{ \limsup_{n\to\infty} \frac{1}{2n+1} \{\ln |(f_\mu^{2n+1})'(y_0)|\}, \right.$$
$$\left. \limsup_{n\to\infty} \frac{1}{2n} \{\ln |(f_\mu^{2n})'(y_0)|\} \right\}. \tag{5.14}$$

Let $g_\mu = f_\mu^2$. Then

$$\limsup_{n\to\infty} \frac{1}{2n} \{\ln |(f_\mu^{2n})'(y_0)|\}$$
$$= \frac{1}{2} \limsup_{n\to\infty} \frac{1}{n} \{\ln |(g_\mu^n)'(y_0)|\}$$
$$= \frac{1}{2} \limsup_{n\to\infty} \frac{1}{n} \{\ln |(g_\mu^n)'(z_0)|\} \qquad \text{(if (5.12) holds)}$$
$$< 0,$$

because we know that z_0 is an attracting fixed point of g_μ^2 and, thus, $|(g_\mu^2)'(z_0)| < 1$. Similarly,

$$\limsup_{n\to\infty} \frac{1}{2n+1} \{\ln |(f_\mu^{2n+1})'(y_0)|\} = \limsup_{n\to\infty} \frac{1}{2n} \{\ln |(f_\mu^{2n})'(y_0)|\} < 0. \qquad \square$$

The argument in Example 5.5 can easily be generalized to the following result, which is left as an exercise.

Exercise 5.6   Prove the following:

"Let $f: I \to I$ be C^1 except at finitely many points. Assume that f has a globally attracting period-n orbit $\mathcal{O} = \{y_0, y_1, \ldots, y_{n-1}\}$ such that $f'(y_j) \neq 0$ for $j = 0, 1, \ldots, n-1$. Then there are infinitely many x_0 such that

$$\lambda(x_0) < 0."$$

$\square$

We thus see that a *negative Lyapunoff exponent* is a sign that there is *no chaos*.

Definition 5.7   Let X and Y be two topological spaces. A map $h: X \to Y$ is said to be a *homeomorphism* if

(i) h is 1-1 and onto;

(ii) h is continuous; and

(iii) h^{-1} is also continuous.   □

Definition 5.8   Let X and Y be two topological spaces and let $f\colon X \to X$ and $g\colon Y \to Y$ be continuous. We say that f and g are *topologically conjugate* if there exists a homeomorphism $h\colon X \to Y$ such that

$$h \circ f(x) = g \circ h(x) \qquad \forall x \in X.$$

$$\begin{array}{ccc} X & \xrightarrow{f} & X \\ h \downarrow & & \downarrow h \\ Y & \longrightarrow & Y \end{array}$$

We call h a *topological conjugacy*.   □

Example 5.9   The quadratic map $f_4(x) = 4x(1-x)$ is topologically conjugate to the roof function $T(x)$ in Example 5.4. Let us verify this by defining

$$h(y) = \sin^2\left(\frac{\pi y}{2}\right), \qquad y \in I = [0, 1].$$

Then

$$h \circ T(y) = \begin{cases} \sin^2\left[\frac{\pi}{2}(2y)\right], & \text{if } 0 \le y \le 1/2; \\ \sin^2\left[\frac{\pi}{2} \cdot 2(1-y)\right], & \text{if } 1/2 < y \le 1, \end{cases}$$
$$= \sin^2(\pi y), \qquad \forall y \in [0, 1].$$

On the other hand,

$$f_4 \circ h(y) = 4\sin^2\left(\frac{\pi y}{2}\right)\left[1 - \sin^2\left(\frac{\pi y}{2}\right)\right]$$
$$= 4\sin^2\left(\frac{\pi y}{2}\right)\cos^2\left(\frac{\pi y}{2}\right)$$
$$= \sin^2(\pi y), \qquad \forall y \in [0, 1].$$

Therefore,

$$h \circ T = f_4 \circ h$$

and

$$f_4 = h \circ T \circ h^{-1}$$
$$f_4^n = (h \circ T \circ h^{-1}) \circ (h \circ T \circ h^{-1}) \circ \cdots \circ (h \circ T \circ h^{-1})$$
$$= h \circ T^n \circ h^{-1}. \quad \square \tag{5.15}$$

Exercise 5.10   For $f_4(x) = 4x(1 - x)$, use computer to calculate the Lyapunoff exponent

$$\lambda(x_0), \quad \text{for} \quad x_0 = \frac{1}{n}, \quad n = 3, 4, 5, \ldots, 10,$$

by approximating

$$\lambda(x_0) \approx \frac{1}{1000} \ln |(f^{1000})'(x_0)|. \qquad\qquad \square$$

Using the topological conjugacy that we established in Example 5.9 above, we can now compute the Lyapunoff exponent of the quadratic map $f_\mu(x) = \mu x(1 - x)$ when $\mu = 4$, as follows.

Example 5.11   For the quadratic map $f_4(x) = 4x(1 - x)$, assume that $x_0 \notin \mathcal{O}^-(1/2) \cup \mathcal{O}^{-1}(0) = \mathcal{O}^{-1}(0)$. We claim that for such x_0,

$$\begin{aligned}\lambda(x_0) &= \text{the Lyapunoff exponent of } x_0 \\ &= \ln 2 > 0.\end{aligned} \tag{5.16}$$

First, note that $h(x) = \sin^2\left(\frac{\pi}{2}x\right)$ is (C^∞) differentiable, so

$$|h'(x)| \le K \qquad \forall x \in [0, 1] \tag{5.17}$$

for some positive constant $K > 0$. Also, note that if x is bounded away from 0 and 1, i.e.,

$$x \in [\delta, 1 - \delta] \quad \text{for some} \quad \delta > 0,$$

then $\sin\left(\frac{\pi x}{2}\right)$ and $\cos\left(\frac{\pi x}{2}\right)$ are bounded away from 0, so

$$|h'(x)| = \left|2 \cdot \frac{\pi}{2} \cos\left(\frac{\pi x}{2}\right) \sin\left(\frac{\pi x}{2}\right)\right| \ge K_\delta, \quad \forall x \in [\delta, 1 - \delta], \tag{5.18}$$

for some constant $K_\delta > 0$.

Now, for $x \notin \mathcal{O}^-(0)$, we have

$$\begin{aligned}\lambda(x_0) &= \limsup_{n \to \infty} \frac{1}{n} [\ln |(f_4^n)'(x_0)|] \\ &= \limsup_{n \to \infty} \frac{1}{n} [\ln |[h \circ T^n \circ h^{-1}]'(x_0)|] \\ &= \limsup_{n \to \infty} \frac{1}{n} [\ln |h'(y_n)| + \ln |(T^n)'(y_0)| + \ln |(h^{-1})'(y_0)|] \\ &\quad (\text{where } y_0 = h^{-1}(x_0), \text{ and } y_n = T^n(y_0)) \\ &\le \lim_{n \to \infty} \frac{1}{n} [\ln K + n \ln 2 + \ln |(h^{-1})'(y_0)|] \\ &\quad (\text{by } (5.17) \text{ and Example 5.4}) \\ &= \ln 2.\end{aligned} \tag{5.19}$$

On the other hand, if we choose a subsequence $\{n_j \mid j = 1, 2, \ldots\} \subseteq \{0, 1, 2, \ldots\}$ such that

$$y_{n_j} = T^{n_j}(y_0) \in [\delta, 1 - \delta],$$

then for $x_0 \notin \mathcal{O}^{-1}(0)$, $x_0 = h(y_0)$, we have

$$
\begin{aligned}
\lambda(x_0) &= \limsup_{n \to \infty} \frac{1}{n}[\ln |(h \circ T^n \circ h^{-1})'(x_0)|] \\
&\geq \limsup_{j \to \infty} \frac{1}{n_j}[\ln |(h \circ T^{n_j} \circ h^{-1})'(x_0)|] \\
&= \lim_{j \to \infty} \frac{1}{n_j}[\ln |h'(y_n)| + n_j \ln 2 + \ln |(h^{-1})'(x_0)|] \\
&\geq \lim_{j \to \infty} \frac{1}{n_j}[\ln K_\delta + n_j \ln 2 + \ln |(h^{-1})'(x_0)|] \quad \text{(by (5.18))} \\
&= \ln 2. \hspace{8cm} (5.20)
\end{aligned}
$$

From (5.19) and (5.20), we conclude (5.16). $\qquad\square$

Next, we study a different kind of chaos involving "fractal" structure. We begin by introducing the concept of a *Cantor set*.

Example 5.12   (The Cantor no-middle-third set) Consider the unit interval $I = [0, 1]$. We first remove the middle-third section of the interval I, i.e.,

$$I - \left(\frac{1}{3}, \frac{2}{3}\right) = [0, 1/3] \cup [2/3, 1].$$

Then for the two closed intervals $[0, 1/3]$ and $[2/3, 1]$, we again remove their respective middle-third sections:

$$
\begin{aligned}
[0, 1/3] - (1/9, 2/9) &= [0, 1/9] \cup [2/9, 1/3], [2/3, 1] - (2/3 + 1/3^2, 2/3 + 2/3^2) \\
&= [2/3, 1] - (7/9, 8/9) = [2/3, 7/9] \cup [8/9, 1].
\end{aligned}
$$

This process is continued indefinitely, as the following diagram shows:

$$0 \qquad 1/9 \quad 2/9 \qquad 1/3 \qquad\quad 2/3 \qquad 7/9 \quad 8/9 \qquad 1$$

Figure 5.6: The process of removing the middle-third open segments of subintervals.

The final outcome of this process is called the *Cantor ternary set*. It has a "*fractal*" structure for if we put it under a microscope, we see that the set looks the same no matter what magnification scale we use, i.e., it is self-similar.

Denote this set by $\mathcal{C}$. Elements in $\mathcal{C}$ are best described from the way $\mathcal{C}$ is created using *ternary representation* of numbers:

$$\mathcal{C} = \left\{ a \in [0, 1] \,|\, a = 0.a_1 a_2 \ldots a_n \ldots = \sum_{n=1}^{\infty} \frac{a_n}{3^n}, a_j \in \{0, 1, 2\} \right.$$
$$\left. \text{satisfying (5.21) below} \right\},$$

where either

$$\text{(i) } \sum_{j=1}^{\infty} |a_j| = \infty, \quad a_j \neq 1 \quad \forall j,$$

or

$$\text{(ii) } \sum_{j=1}^{\infty} |a_j| < \infty, \text{ i.e., } a = 0.a_1 a_2 \ldots a_n 000 \ldots, \quad a_n \in \{1, 2\};$$
$$a_j \neq 1 \,\forall j: \; 1 \leq j \leq n - 1.$$

$$\tag{5.21}$$

$\square$

Definition 5.13 Let $S \subseteq \mathbb{R}^N$. A point $x_0 \in \mathbb{R}^N$ is said to be an *accumulation point* of S if every open set containing x_0 also contains at least one point of $S \backslash \{x_0\}$. $\square$

Definition 5.14 A set S in $\mathbb{R}$ is said to be *totally disconnected* if S does not contain any interval. $\square$

Definition 5.15 A set $S \subseteq \mathbb{R}^n$ is said to be a *perfect set* if every point is an accumulation point of S. $\square$

Theorem 5.16 The Cantor ternary set $\mathcal{C}$ is closed, totally disconnected and perfect.

Proof. The set $\mathcal{C}$ is closed because

$$\mathcal{C} = I - \bigcup_{j=1}^{\infty} I_j \tag{5.22}$$

where I_j's are the middle-third open intervals that are removed in the construction process of $\mathcal{C}$. Since $\bigcup_{j=1}^{\infty} I_j$ is a union of open intervals and any union of open intervals is open, $\mathcal{C}$ as given by (5.22) is closed.

Assume that $\mathcal{C}$ is not totally disconnected. Then $\mathcal{C}$ contains some intervals. By the construction process of $\mathcal{C}$, this interval will have a subinterval whose middle-third will be removed. So this middle-third interval does not belong to $\mathcal{C}$, a contradiction.

To show that $\mathcal{C}$ is perfect, we need to show that every point $a \in \mathcal{C}$ is an accumulation point. We use (5.21). If

(i) $a = 0.a_1a_2 \ldots a_n \ldots$ such that $\sum_{j=1}^{\infty} |a_j| = \infty$ and $a_j \neq 1 \ \forall j$, then a is an accumulation point of

$$\{y \mid y = 0.a_1a_2 \ldots a_m, \ \text{for } m = 1, 2, \ldots, k, k+1, \ldots\} \subseteq \mathcal{C};$$

(ii) $a = 0.a_1a_2 \ldots a_n$ where $a_n \in \{1, 2\}$ and $a_j \neq 1$ for $j = 1, 2, \ldots, n-1$, then a is an accumulation point of

$$\{y \mid y = 0.a_1a_2 \ldots a_{n-1}b_nb_{n+1} \ldots b_{n+k}, \ \text{for } k = 0, 1, 2, \ldots,$$
$$b_j = 2 \text{ for } j = n, n+1, \ldots, n+k\}. \tag{5.23}$$

$\square$

NOTES FOR CHAPTER 5

The term "homoclinic" or "homoclinicity" cannot not even be found in most major English dictionaries. Basically, here it means "of the same orbit", while "heteroclinic" is its antonym, meaning "of the different orbit".

Homoclinicity is a global concept. In this chapter, we have only discussed it for interval maps. A point x_0 in a neighborhood of a repelling fixed point p will move away from p after a few iterations at first. But some actually return exactly to p. This is *totally astonishing*. We did not discuss homoclinic orbits of *higher-dimensional maps* in this book. They may have highly complex behaviors. We refer the readers to Robinson [58] and Wiggins [69], for example.

The definition of Lyapunoff exponents for smooth interval maps can be traced back to Lyapunoff's dissertation [50] in 1907. We did not discuss that for multidimensional diffeomorphisms, but they can easily be found in many dynamical systems books. Lyapunoff exponents provide an easily computable quantity, using commercial or open source algorithms and softwares (see [27], [32], [70], Maple, Matlab, etc.). For example, the Maple software computer codes written by Rucklidge [60] for computing the Lyapunoff exponents of the quadratic map consists of only 21 lines. Many papers in the engineering literature, where rigorous proofs are not required, claim that the systems under study are chaotic once the authors are able to compute that the Lyapunoff exponents are greater than one.

CHAPTER 6

Symbolic Dynamics, Conjugacy and Shift Invariant Sets

6.1 THE ITINERARY OF AN ORBIT

Let us now consider the quadratic map $f_\mu(x) = \mu x(1-x)$ for $\mu > 4$. When $\mu > 4$, $I = [0, 1]$ is no longer invariant under f_μ. Part of I will be mapped outside I, as can be seen in Fig. 6.1. Also, from Fig. 6.1, we see that there are two subintervals I_0 and I_1 of I, such that

$$f_\mu(I_0) = I, \qquad f_\mu(I_1) = I. \tag{6.1}$$

Apply f_μ^2 to I_0 and I_1, respectively, we see that I_0 (resp., I_1) will have a subinterval mapped out of I. So we remove that (open) subinterval from I_0 (resp., I_1). This process can be continued indefinitely. We see that it is analogous to the process of constructing the Cantor ternary set $\mathcal{C}$ in Chapter 5. The outcome of this process, which is obtained by removing all the points which are eventually mapped out of I, is denoted as

$$\Lambda = \{x \in I \mid f_\mu^n(x) \in I, \quad \forall n = 0, 1, 2, \ldots\}. \tag{6.2}$$

This Λ is invariant under f_μ:

$$f_\mu : \ \Lambda \longrightarrow \Lambda.$$

Also, if $x \in I \backslash \Lambda$, then $\lim_{n \to \infty} f_\mu^n(x) = -\infty$.

For each $x \in \Lambda$, define its *itinerary* or *symbol* as follows:

$$\begin{aligned}
S(x) &= \text{itinerary of } x \\
&= (s_0 s_1 s_2 \ldots s_n \ldots), \qquad s_j \in \{0, 1\} \quad \forall j, \\
s_j &= \begin{cases} 0 & f_\mu^j(x) \in I_0; \\ & \text{if} \\ 1 & f_\mu^j(x) \in I_1. \end{cases}
\end{aligned} \tag{6.3}$$

We collect all such binary strings (6.3) together and define

$$\begin{aligned}
\sum\nolimits_2 &= \text{the space of all binary symbols} \\
&= \{s \mid s = (s_0 s_1 \ldots s_n \ldots), \text{ where } s_j \in \{0, 1\}, j = 0, 1, 2, \ldots\}.
\end{aligned}$$

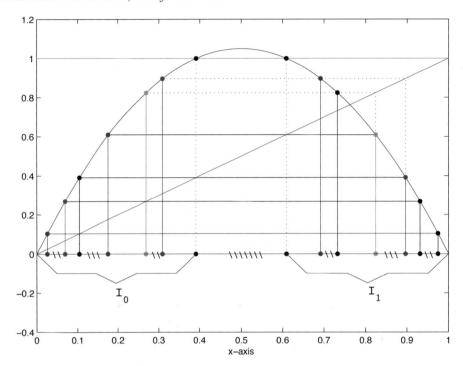

Figure 6.1: The graph of $f_\mu(x) = \mu x(1-x)$ has a portion that exceeds the height of 1. Here, we show that the interval $I = [0, 1]$ has $1 + 2 + 2^2 = 7$ open segments removed. (6.1).

For any $s, t \in \sum_2$, where

$$s = (s_0 s_1 s_2 \ldots s_n \ldots), \qquad t = (t_0 t_1 t_2 \ldots t_n \ldots),$$

define

$$d(s, t) \equiv \text{the distance between } s \text{ and } t$$

$$= \sum_{j=0}^{\infty} \frac{|s_j - t_j|}{2^j}.$$

Then for any $s, t, u \in \sum_2$, we can easily verify that the following triangle inequality is satisfied:

$$d(s, t) \leq d(s, u) + d(u, t). \tag{6.4}$$

We say the $(\sum_2, d)$ forms a metric space, and d is a metric on $\sum_2$.

On $\sum_2$, we now define a "left-shift" map

$$\sigma : \sum_2 \longrightarrow \sum_2,$$
$$\sigma(s) = \sigma(s_0 s_1 \ldots s_n \ldots) = (s_1 s_2 \ldots s_n \ldots).$$

This map σ looks simple and innocent. Yet it is totally surprising to learn that σ is actually chaotic!

6.2 PROPERTIES OF THE SHIFT MAP σ

We prove some basic properties of σ in this section.

Lemma 6.1   Let $s, \tilde{s} \in \sum_2$, where

$$s = (s_0 s_1 \ldots s_n s_{n+1} s_{n+2} \ldots), \qquad \tilde{s} = (s_0 s_1 \ldots s_n \tilde{s}_{n+1} \tilde{s}_{n+2} \ldots),$$

i.e., s and $\tilde{s}$ are identical up to the first $n+1$ bits. Then

$$d(s, \tilde{s}) \leq 2^{-n}.$$

Conversely, if $s, \tilde{s} \in \sum_2$ such that $d(s, \tilde{s}) < 2^{-n}$, then s and $\tilde{s}$ must agree up to at least the first $n+1$ bits.

Proof. We have

$$d(s, \tilde{s}) = \sum_{j=0}^{n} \frac{|s_j - s_j|}{2^j} + \sum_{j=n+1}^{\infty} \frac{|s_j - \tilde{s}_j|}{2^j}$$

$$\leq \sum_{j=n+1}^{\infty} \frac{1}{2^j} = 2^{-n}.$$

The proof of the converse follows also in a similar way. □

Theorem 6.2   The map $\sigma : \sum_2 \longrightarrow \sum_2$ is continuous with respect to the metric (6.4).

Proof. We need to prove the following: for any given $\tilde{s} \in \sum_2$, for any $\varepsilon > 0$, there exists a $\delta > 0$ such that if $d(\tilde{s}, t) < \delta$, then $d(\sigma(\tilde{s}), \sigma(t)) < \varepsilon$.

Find a positive integer n sufficiently large such that

$$\frac{1}{2^{n-1}} < \varepsilon, \quad \text{and choose} \quad \delta = \frac{1}{2^n}.$$

Write

$$\tilde{s} = (s_0 s_1 \ldots s_n \ldots).$$

Consider any $t \in \sum_2$ such that $d(\tilde{s}, t) < \delta = 2^{-n}$. Then by Lemma 6.1, we know that

$$t = (s_0 s_1 \ldots s_n t_{n+1} t_{n+2} \ldots),$$

i.e., t must agree with $\tilde{s}$ in the first $n + 1$ bits. Thus,

$$\sigma(\tilde{s}) = (s_1 s_2 \ldots s_n s_{n+1} s_{n+2} \ldots), \quad \sigma(t) = (s_1 s_2 \ldots s_n t_{n+1} t_{n+2} \ldots)$$

and

$$d(\sigma(\tilde{s}), \sigma(t)) = \sum_{j=1}^{n} \frac{|s_j - s_j|}{2^{j-1}} + \sum_{j=n+1}^{\infty} \frac{|s_j - t_j|}{2^{j-1}}$$

$$= \sum_{j=n+1}^{\infty} \frac{|s_j - t_j|}{2^{j-1}} \leq \sum_{j=n+1}^{\infty} \frac{1}{2^{j-1}} = 2^{-(n-1)} < \varepsilon. \tag{6.5}$$

$\square$

Let A and B be two sets in a metric space X. We say that A is *dense* in B if $A \subset B$ and for every $y \in B$, there exists a sequence $\{x_n\}_{n=1}^{\infty} \subseteq A$ such that $y = \lim_{n \to \infty} x_n$. Thus, $\bar{A} = B$, where $\bar{A}$ is the closure of A.

Theorem 6.3   Consider $\sigma : \sum_2 \longrightarrow \sum_2$. Denote

$$\mathrm{Per}_n(\sigma) = \text{the set of all points in } \sum_2 \text{ with period less than or equal to } n$$

and

$$\mathrm{Per}(\sigma) = \bigcup_{n=1}^{\infty} \mathrm{Per}_n(\sigma).$$

Then

(i) The cardinality of $\mathrm{Per}_n(\sigma)$ is 2^n; and

(ii) $\mathrm{Per}(\sigma)$ is dense in $\sum_2$.

Proof.   (i) If $s \in \mathrm{Per}_n(\sigma)$, then it is easy to see that

$$s = (\overline{s_0 s_1 \ldots s_{n-1}}) \equiv (s_0 s_1 s_2 \ldots s_{n-1} s_0 s_1 \ldots s_{n-1} s_0 s_1 \ldots s_{n-1} \ldots),$$

$$\overbrace{}^{n \text{ times}}$$

and vice versa. There are $2 \times 2 \times \cdots \times 2 = 2^n$ different combinations of $\mathrm{Per}_n(\sigma)$. Therefore, the cardinality of $\mathrm{Per}_n(\sigma)$ is 2^n.

(ii) For any given

$$\tilde{s} = (\tilde{s}_0 \tilde{s}_1 \tilde{s}_2 \ldots \tilde{s}_n \tilde{s}_{n+1} \ldots),$$

define

$$\tilde{s}_k = (\overline{\tilde{s}_0 \tilde{s}_1 \ldots \tilde{s}_{k-1}}), \quad \text{for } k = 1, 2, 3, \ldots .$$

Then $\tilde{s}_k \in \text{Per}_k(\sigma)$. By Lemma 6.1, we have

$$d(\tilde{s}, \tilde{s}_k) \leq \frac{1}{2^{k-1}} \longrightarrow 0 \quad \text{as} \quad k \to \infty.$$

Therefore, $\text{Per}(\sigma)$ is dense in $\sum_2$.

$\square$

Theorem 6.4   There exists a $\tau_0 \in \sum_2$ such that $\overline{\mathcal{O}(\tau_0)} = \sum_2$.

Proof. We construct τ_0 as follows:

$$\tau_0 = \Big(\underbrace{0 \; 1}_{\text{block 1}} : \underbrace{00 \; 01 \; 10 \; 11}_{\text{block 2}} : \underbrace{000 \; 001 \; 010 \; 011 \; 100 \; 101 \; 110 \; 111}_{\text{block 3}} : \underbrace{\cdots}_{\text{block 4}}$$

$$\cdots : \underbrace{\cdots \cdots}_{\text{block } n} : \underbrace{\cdots \cdots \cdots}_{\text{block } n+1} : \cdots \cdots \Big)$$

where the n block consists of all n-bit strings arranged sequentially in ascending order.

For any $s \in \sum_2$, write $s = (s_0 s_1 \ldots s_k \ldots)$. The first k bits of s, i.e., $s_0 s_1 \ldots s_{k-1}$, appear somewhere in the k-th block of τ_0. Thus, there exists some $n(k)$ such that

$$\sigma^{n(k)}(\tau_0) = (s_0 s_1 \ldots s_{k-1} t_k t_{k+1} \ldots)$$

where $t_{k+j} \in \{0, 1\}$ for $j = 0, 1, 2, \ldots$. Thus, by Lemma 6.1, we have

$$d(s, \sigma^{n(k)}(\tau_0)) \leq 1/2^{k-1}.$$

Since k is arbitrary, we see that

$$\lim_{k \to \infty} d(s, \sigma^{n(k)}(\tau_0)) = 0.$$

Therefore, $\overline{\mathcal{O}(\tau_0)} = \sum_2$.

$\square$

We now introduce the important concept of *topological transitivity* in the following.

Definition 6.5   Let X be a topological space and $f \colon X \to X$ be continuous. We say that f is *topologically transitive* if for every pair of nonempty open sets U and V of X, there exists an $n > 0$ such that $f^n(U) \cap V \neq \emptyset$.

$\square$

The topological transitivity of a map causes the "mixing" of any two open sets by that map.

Exercise 6.6   Let X be a compact metric space and $f: X \to X$ be continuous and onto. Show the following statements are equivalent:

(i) f is topologically transitive.

(ii) If Λ is a closed subset of X and $f(\Lambda) \subset \Lambda$, then either $\Lambda = X$ or Λ is nowhere dense in X.

(iii) If E is a open subset of X and $f^{-1}(E) \subset E$, then either $E = \emptyset$ or E is everywhere dense in X. □

Theorem 6.7   The map $\sigma: \sum_2 \longrightarrow \sum_2$ is topologically transitive.

Proof. Let U and V be any two nonempty open sets in $\sum_2$. Since $\overline{\mathcal{O}(\tau_0)} = \sum_2$ by Theorem 6.4, we have

$$\mathcal{O}(\tau_0) \cap U \neq \emptyset.$$

Thus, there exists an $\alpha_0 \in \mathcal{O}(\tau_0) \cap U$. We have

$$\alpha_0 = \sigma^{\ell_1}(\tau_0) \quad \text{for some positive integer} \quad \ell_1.$$

But then from the proof of Theorem 6.4, we easily see that

$$\overline{\mathcal{O}(\alpha_0)} = \sum_2.$$

Thus, $\mathcal{O}(\alpha_0) \cap V \neq \emptyset$. Choose $\beta_0 \in \mathcal{O}(\alpha_0) \cap V$. Then

$$\beta_0 = \sigma^{\ell_2}(\alpha_0) \in V.$$

Therefore, $\beta_0 \in \sigma^{\ell_2}(U) \cap V \neq \emptyset$. □

An important property for many genuinely nonlinear maps is that of *sensitive dependence* on initial data.

Definition 6.8   Let (X, d) be a metric space and $f: X \to X$ be continuous. We say that f has *sensitive dependence on initial data* if there exists a $\delta > 0$ such that for every $x \in X$ and for every open neighborhood $\mathcal{N}(x)$ of x, there exist a $y \in \mathcal{N}(x)$ and an n (depending on y) such that

$$d(f^n(x), f^n(y)) \geq \delta.$$

Here δ is called the sensitivity constant of f. □

Theorem 6.9   The map $\sigma: \sum_2 \longrightarrow \sum_2$ is sensitively dependent on initial data.

Proof. Let $\eta_0 = (s_0 s_1 \ldots s_n \ldots) \in \Sigma_2$ and let $\mathcal{N}(\eta_0)$ be an open neighborhood of η_0 in Σ_2. Define, for a positive integer m_0,

$$A = \left\{ s \in \sum_2 \,\bigg|\, s \text{ agrees with } \eta_0 \text{ for the first } m_0 \text{ bits} \right\}.$$

If $\zeta_0 \in A$, then

$$d(\eta_0, \zeta_0) \leq \frac{1}{2^{m_0}}, \quad \text{by Lemma 6.1.}$$

We choose m_0 sufficiently large such that $A \subseteq \mathcal{N}(\eta_0)$. Also, choose $\delta = 1$ and

$$\eta_1 = (s_0 s_1 \ldots s_{m_0} t_{m_0+1} t_{m_0+2} \ldots),$$

where

$$t_{m_0+1} = s_{m_0+1} + 1 \pmod{2}$$

but

$$t_{m_0+j} \in \{0, 1\} \quad \text{is arbitrary for} \quad j > 1.$$

Then $\eta_1 \in A \subseteq \mathcal{N}(\eta_0)$ and

$$d(\sigma^{m_0+1}(\eta_0), \sigma^{m_0+1}(\eta_1)) = d((s_{m_0+1} s_{m_0+2} \ldots s_{m_0+k} \ldots),$$
$$(t_{m_0+1} t_{m_0+2} \ldots t_{m_0+k} \ldots)) \geq 1 = \delta. \tag{6.6}$$

$\square$

Summarizing Theorems 6.3, 6.7 and 6.9, we see that the shift map $\sigma : \sum_2 \longrightarrow \sum_2$ satisfies the following three properties:

(i)   The set of all periodic points is dense;
(ii) It is topologically transitive; and   (6.7)
(iii) It has sensitive dependence on initial data.

Properties (i), (ii) and (iii) above are known to be (individually) independent of each other. R.L. Devaney [20] used these three important properties to define chaos as follows:

"Let (X, d) be a metric space and $f : X \to X$ be continuous. We say that the map f is *chaotic* on X if f satisfies (i)–(iii) above."

However, a paper by Banks, Brooks, Cairns and Stacey [3] points out that conditions (i) and (ii) actually imply (iii). Thus, condition (iii) is redundant in Devaney's definition. See Theorem 6.26 later.

Exercise 6.10   For the quadratic map $f_\mu(x) = \mu x(1 - x)$, let $\mu > 2 + \sqrt{5}$. Define

$$I_0 = \{x \in [0, 1] \mid 0 \leq x < 1/2, f_\mu(x) \leq 1\},$$
$$I_1 = \{x \in [0, 1] \mid 1/2 \leq x \leq 1, f_\mu(x) \leq 1\}.$$

Prove that $|f'_\mu(x)| \geq 1 + \delta$ for some $\delta > 0 \; \forall x \in I_0 \cup I_1$. □

Let us explore the properties of $S: \Lambda \longrightarrow \sum_2$ and $\sigma: \sum_2 \longrightarrow \sum_2$ and their relationship.

Theorem 6.11   Assume that $\mu > 2 + \sqrt{5}$. Then we have

$$S \circ f_\mu = \sigma \circ S \tag{6.8}$$

as indicated in the commutative diagram in Fig. 6.2.

$$
\begin{array}{ccc}
\Lambda & \xrightarrow{\;f_\mu\;} & \Lambda \\
S \downarrow & & \downarrow S \\
\sum_2 & \xrightarrow{\;\sigma\;} & \sum_2
\end{array}
$$

Figure 6.2: Commutative diagram.

Proof. Let $x \in \Lambda$ such that

$$S(x) = \text{the itinerary of } x = (s_0 s_1 \ldots s_n \ldots).$$

Denote $y = f_\mu(x) \in \Lambda$, with $S(y) = (t_0 t_1 \ldots t_n \ldots)$. Then by the definition of $S(x)$ and $S(y)$, we have

$$
\begin{aligned}
x \in I_{s_0}, &\quad f_\mu(x) \in I_{s_1}, \quad f^2_\mu(x) \in I_{s_2}, \ldots, f^n_\mu(x) \in I_{s_n}, \ldots, \\
y \in I_{t_0}, &\quad f_\mu(x) \in I_{t_1}, \quad f^2_\mu(y) \in I_{t_2}, \ldots, f^n_\mu(y) \in I_{t_n}, \ldots.
\end{aligned}
$$

But $f^n_\mu(y) = f^n_\mu(f_\mu(x)) = f^{n+1}_\mu(x)$ for any $n = 0, 1, 2, \ldots$. Therefore,

$$
f^n_\mu(y) \in I_{t_n} \text{ and } f^n_\mu(y) = f^{n+1}_\mu(x) \in I_{s_{n+1}}, \text{ implying } t_n = s_{n+1}, \\
\forall n = 0, 1, 2, \ldots,
$$

i.e., $S(y) = (s_1 s_2 \ldots s_n s_{n+1} \ldots) = S(f_\mu(x)) = \sigma(s_0 s_1 \ldots s_n \ldots) = \sigma(S(x))$. □

Theorem 6.12   Assume that $\mu > 2 + \sqrt{5}$. Then $S: \Lambda \to \sum_2$ is a homeomorphism.

Proof. We need to prove that S is 1-1, onto, continuous, and that S^{-1} is also continuous.

(i) 1-1: Let $x, y \in \Lambda$ such that $S(x) = S(y)$. This means that x and y have the same itinerary, i.e., $f_\mu^n(x)$ and $f_\mu^n(y)$ belong to the same I_j for $j \in \{0, 1\}$ for any $n = 0, 1, 2, \ldots$. Since f_μ is monotonic on both I_0 and I_1, every point $z \in (x, y)$ satisfy the same itinerary as that of x and y. Therefore, f_μ^n maps the closed interval $[x, y] \equiv J$ to either I_0 or I_1, for any $n = 0, 1, 2, \ldots$. But, by the fact that $\mu > 2 + \sqrt{5}$, we have

$$|f_\mu'(x)| \geq 1 + \delta \quad \text{for some} \quad \delta > 0, \quad \forall x \in I_0 \cup I_1.$$

Using the mean value theorem, we therefore have

$$\text{length of } f_\mu^n(J) \geq (1 + \delta)^n \cdot (y - x) \longrightarrow \infty \text{ as } n \to \infty,$$

a contradiction.

(ii) Onto: Choose any $s = (s_0 s_1 \ldots s_n \ldots) \in \sum_2$. We want to find an $x \in \Lambda$ such that $S(x) = s$. Define, for any $n = 0, 1, 2, \ldots$

$$I_{s_0 s_1 \ldots s_n} = \{x \in I = [0, 1] \mid x \in I_{s_0} f_\mu(x) \in I_{s_1}, \ldots, f_\mu^n(x) \in I_{s_n}\} \tag{6.9}$$
$$= I_{s_0} \cap f_\mu^{-1}(I_{s_1}) \cap f_\mu^{-2}(I_{s_2}) \cap \cdots \cap f_\mu^{-n}(I_{s_n}).$$

Then

$$I_{s_0 s_1 \ldots s_n} = [I_{s_0} \cap f_\mu^{-1}(I_1) \cap \cdots \cap f_\mu^{-(n-1)}(I_{s_{n-1}})] \cap f_\mu^{-n}(I_{s_n})$$
$$= I_{s_0 s_1 \ldots s_{n-1}} \cap f_\mu^{-n}(I_{s_n}).$$

Thus, $I_{s_0 s_1 \ldots s_n} \subset I_{s_0 s_1 \ldots s_{n-1}}$ and the closed intervals $I_{s_0 s_1 \ldots s_n}$ form a nested sequence of nonempty closed sets. From topology, we know that

$$\bigcap_{n=0}^{\infty} I_{s_0 s_1 \ldots s_n} \neq \emptyset.$$

Therefore, there exists some $x \in \bigcap_{n=0}^{\infty} I_{s_0 s_1 \ldots s_n}$. By definition, $x \in \Lambda$, and $S(x) = (s_0 s_1 \ldots s_n \ldots)$. This x is actually unique by part (i) of the proof.

(iii) S is continuous: For any given $\varepsilon > 0$, at any point $x \in \Lambda$, we want to choose a $\delta > 0$ such that

$$d(S(x), S(y)) < \varepsilon \quad \text{provided that} \quad |x - y| < \delta, \quad \forall y \in \Lambda.$$

First, let n be a positive integer such that $2^{-n} < \varepsilon$. Let $I_{s_0 \ldots s_n}$ be defined as in (6.9) such that $x \in I_{s_0 s_1 \ldots s_n}$. Choose $\delta > 0$ sufficiently small such that if $|y - x| < \delta$ and $y \in \Lambda$, we have $y \in I_{s_0 s_1 \ldots s_n}$. Then

$$S(y) = (s_0 s_1 \ldots s_n t_{n+1} t_{n+2} \ldots t_{n+k} \ldots)$$

where $t_{n+k} \in \{0, 1\}$ for any $k = 1, 2, \ldots$. This gives

$$d(S(x), S(y)) \leq \frac{1}{2^n} < \varepsilon.$$

(iv) S^{-1} is continuous from $\sum_n$ to Λ: This is left as an exercise.   □

□

Remark 6.13   Theorem 6.12 is also true as long as $\mu > 4$, but the proof is somewhat more involved. In the proof of Theorem 6.12, we have utilized the property that

$$|f'_\mu(x)| \geq 1 + \delta \qquad \forall x \in I_0 \cup I_1, \qquad (\mu > 2 + \sqrt{5}) \qquad (6.10)$$

which is not true if $4 < \mu \leq 2 + \sqrt{5}$. However, instead of (6.10), we can utilize the property that for any μ such that $4 < \mu \leq 2 + \sqrt{5}$, we have

$$|(f^n_\mu)'(x)| \geq 1 + \delta \qquad \forall x \in I_0 \cup I_1,$$

for some positive integer n. Thus, the proof goes through.   □

Since $S : \Lambda \to \sum_2$ is a homeomorphism, by Theorem 6.11, we can write

$$f_\mu = S^{-1} \circ \sigma \circ S.$$

Therefore, many useful properties of σ pass on to f_μ, as stated below.

Corollary 6.14   Let $f_\mu(x) = \mu x(1 - x)$ with $\mu > 4$. Then the map $f_\mu : \Lambda \to \Lambda$ has the following properties:

(i) The cardinality of $\mathrm{Per}_n(f_\mu)$ is 2^n.

(ii) $\mathrm{Per}(f_\mu)$ is dense in Λ.

(iii) The map f_μ has a dense orbit, i.e., there exists an $x_0 \in \Lambda$ such that $\mathcal{O}(x_0)$ is dense in Λ.   □

6.3   SYMBOLIC DYNAMICAL SYSTEMS $\sum_k$ AND $\sum_k^+$

Consider the set consisting of k symbols,

$$S(k) = \{0, 1, \ldots, k - 1\}. \qquad (6.11)$$

Endowed with the discrete topology, this set becomes a topological space. All the subsets in $S(k)$ are open. The set $S(k)$ is metrizable, and a compatible metric is

$$\delta(a, b) = \begin{cases} 1, & \text{if } a \neq b, \\ 0, & \text{if } a = b. \end{cases} \qquad (6.12)$$

Obviously, $S(k)$ is compact, so the topological products

$$\sum_k = \prod_{j=-\infty}^{+\infty} S_j, \qquad S_j = S(k),$$

or

$$\sum_k^+ = \prod_{j=0}^{+\infty} S_j, \qquad S_j = S(k),$$

are also compact spaces by the well known Tychonov Theorem [26]. $\sum_k$ and $\sum_k^+$ are called, respectively, two-side and one-side symbol spaces with k symbols. Any element s in $\sum_k$ is a two-sided symbol sequence

$$s = (\ldots, s_{-2}, s_{-1}; s_0, s_1, s_2, \ldots),$$

and a one-side symbol sequence in $\sum_k^+$

$$s = (s_0, s_1, s_2, \ldots),$$

respectively. $\sum_k$ and $\sum_k^+$ are metrizable, A usual metric is, for two-side case,

$$d(s,t) = \sum_{j=-\infty}^{+\infty} \frac{\delta(s_j, t_j)}{2^{|j|}}, \tag{6.13}$$

for

$$\begin{cases} s = (\ldots, s_{-2}, s_{-1}; s_0, s_1, s_2, \ldots), \\ t = (\ldots, t_{-2}, t_{-1}; t_0, t_1, t_2, \ldots), \end{cases} \tag{6.14}$$

and for the one-side case

$$d(s,t) = \sum_{j=0}^{+\infty} \frac{\delta(s_j, t_j)}{2^j}, \tag{6.15}$$

for

$$s = (s_0, s_1, s_2, \ldots),$$
$$t = (t_0, t_1, t_2, \ldots).$$

Exercise 6.15   Prove that $d(\cdot, \cdot)$ defined by (6.13) and (6.15) is a metric on, respectively, $\sum_k$ and $\sum_k^+$. Furthermore, $(\sum_k, d)$ and $(\sum_k^+, d)$ are compact metric spaces.

Exercise 6.16   In lieu of (6.13), define

$$d(s,t) = \sum_{j=-\infty}^{\infty} \frac{1}{2^{|j|}} \frac{|s_j - t_j|}{1 + |s_j - t_j|}.$$

Prove that $d(\cdot, \cdot)$ is a metric on $\sum_k$ equivalent to (6.15). □

On $\sum_k$, we now define a "left-shift" map

$$\sigma: \sum_k \longrightarrow \sum_k$$

$$\sigma(s) = \sigma(\ldots, s_{-2}, s_{-1}, \dot{s}_0, s_1, s_2, \ldots) = (\ldots, s_{-2}, s_{-1}, s_0, \dot{s}_1, s_2, \ldots).$$

Similarly, on $\sum_k^+$, we define a "left-shift" map

$$\sigma^+: \sum_k^+ \longrightarrow \sum_k^+$$

$$\sigma^+(s) = \sigma(s_0, s_1, s_2, \ldots) = (s_1, s_2, \ldots).$$

Lemma 6.17

(1) σ is a homeomorphism from $\sum_k$ to itself.

(2) σ^+ is continuous from $\sum_k^+$ onto itself.

Proof. Let s and t be given by (6.14). We easily verify the following:

$$d(s, t) < 2^{-n} \Rightarrow s_j = t_j, \quad j = 0, \pm 1, \pm 2, \ldots, \pm n \Rightarrow d(s, t) \le 2^{-(n-1)}. \tag{6.16}$$

For the continuity of σ, we need to prove the following: for any given $s \in \sum_k$, for any $\varepsilon > 0$, there exists a $\delta > 0$ such that if $d(s, t) < \delta$, then $d(\sigma(s), \sigma(t)) < \varepsilon$.

Find a positive integer n sufficiently large such that $2^{-(n-1)} < \varepsilon$, and choose $\delta = 2^{-(n+1)}$. When $d(s, t) < \delta, s_j = t_j, j = 0, \pm 1, \pm 2, \ldots, \pm(n + 1)$ by (6.16), and so $(\sigma(s))_j = (\sigma(t))_j, j = 0, \pm 1, \pm 2, \ldots, \pm n$. Again by (6.16), we have

$$d(\sigma(s), \sigma(t)) \le 2^{-(n-1)} < \varepsilon.$$

Obviously, the map σ is 1-1 and onto. The continuity of its inverse σ^{-1} can be proved similarly. Therefore, σ is a homeomorphism.

With a similar argument, we prove (2). □

We now have the dynamical systems $(\sum_k, \sigma)$ and $(\sum_k^+, \sigma^+)$ which are called the two-sided shift and one-sided shift, respectively. The dynamics between them are very similar. For simplicity, in the following, we just discuss the dynamics of the one-sided shift. Actually, nearly all the results on $(\sum_k^+, \sigma^+)$ are also true for the two-sided shift $(\sum_k, \sigma)$.

6.4   THE DYNAMICS OF $(\sum_k^+, \sigma^+)$ AND CHAOS

Throughout this subsection, we assume $k \geq 2$, i.e., the symbol space $\sum_k^+$ with the number of symbols not fewer than 2.

Theorem 6.18   Consider $(\sum_k^+, \sigma^+)$. Denote

$$\text{Per}_n(\sigma^+) = \text{the set of all points in } \sum_k^+ \text{ with period less than or equal to } n,$$

and

$$\text{Per}(\sigma^+) = \bigcup_{n=1}^{+\infty} \text{Per}_n(\sigma^+).$$

Then

(1) the cardinality of $\text{Per}_n(\sigma^+)$ is k^n; and

(2) $\text{Per}(\sigma^+)$ is dense in $\sum_k^+$.

Proof.   (1) If $s \in \text{Per}_n(\sigma^+)$, then it is easy to show (Exercise!) that

$$s = (\overline{s_0 s_1 \ldots s_{n-1}}) \equiv (s_0 s_1 \ldots s_{n-1} s_0 s_1 \ldots s_{n-1} \ldots),$$

and vice verse. There are $\overbrace{k \times k \times \cdots k}^{n \text{ times}} = k^n$ different combinations of $\text{Per}_n(\sigma^+)$. Therefore, the cardinality of $\text{Per}_n(\sigma^+)$ is k^n.

(2) For any given

$$s = (s_0 s_1 \ldots s_n s_{n+1} \ldots) \in \sum_k^+,$$

define

$$\tilde{s}^n = (\overline{s_0 s_1 \ldots s_{n-1}}), \qquad n = 1, 2, \ldots .$$

Then $\tilde{s}^n \in \text{Per}_n(\sigma^+)$. By (6.16), we have

$$d(s, \tilde{s}^n) \leq \frac{1}{2^{n-1}} \to 0 \qquad (n \to \infty).$$

Therefore, $\text{Per}(\sigma^+)$ is dense in $\sum_k^+$.

$\square$

Theorem 6.19   The shift map σ^+ is topological transitive. That is, there exists an $s \in \sum_k^+$ such that

$$\overline{orb(s)} = \sum_k^+,$$

where $orb(s)$ denotes the orbit $\{s, \sigma^+(s), \ldots, (\sigma^+)^n(s), \ldots\}$ starting from the point s.

Proof. For any positive integer n, taking n symbols from the k symbols each time (the symbols are allowed to be repeated), we have a total of k^n different order-n blocks. Such k^n many n-blocks are arranged in any order to form an nk^n-block, which is denoted by P_{nk^n}. Construct s as

$$s = (P_k P_{2k^2} \ldots P_{nk^n} \ldots) \in \sum_k^+.$$

We claim that

$$\overline{orb(s)} = \sum_k^+.$$

For any $t \in \sum_k^+$ and for any $n \geq 1$, it follows from the construction of s that there exists an $\ell_n \geq 1$ such that $(\sigma^+)^{\ell_n}(s)$ and t are identical up to the first $n + 1$ places. Then by (6.16), we have

$$d((\sigma^+)^{\ell_n}(s), t) \leq \frac{1}{2^{n-1}} \to 0, \qquad (n \to \infty).$$

Therefore, $\overline{orb(s)} = \sum_k^+$.   □

From Theorem 6.19, the one side shift $(\sum_k^+, \sigma^+)$ is topologically transitive. There exists a point $s = (s_0 s_1 \ldots) \in \sum_k^+$ such that the orbit $\{(\sigma^+)^n(s) \colon n = 0, 1, \ldots\}$ is dense in $\sum_k^+$. Denote by $\sum_k^*$ the set of all such points. The following Proposition shows that the set $\sum_k^*$ is almost equal to the whole symbol space $\sum_k^+$. See [59, Theorem 14.3.20].

Proposition 6.20   Let $\{p_0, \ldots, p_{k-1}\}$ be a system of weights, with $0 < p_i < 1$ for all $0 \leq i \leq k - 1$ and $p_0 + \cdots + p_{k-1} = 1$.

(i) Then, there is a probability measure μ_p defined on $\sum_k^+$ that is invariant by σ^+ and that satisfies the following: given $a_0, \ldots, a_k \in \{0, 1, \ldots k - 1\}$, then

$$\mu_p(s \colon s_i = a_0) = p_{a_0},$$

and

$$\mu_p(s \colon s_i = a_i, i = 0, \ldots, n) = \prod_{i=0}^{n} p_{a_i}.$$

(ii) $\mu_p(\sum_k^*) = 1.$ □

Let X be a compact metric space and $f: X \to X$ be continuous. Define

$$f \times f: X \times X \to X \times X,$$
$$(f \times f)(x, y) = (f(x), f(y)).$$

Then we obtain a topological system $(X \times X, f \times f)$ with $f \times f$ being continuous on the topological product space $X \times X$.

Definition 6.21   Let f be continuous from a compact metric space X into itself. We say that f is *weakly topologically mixing* if $f \times f$ is *topologically transitive*.   □

It is easy to see from the definition that weakly topologically mixing implies topological transitivity.

Theorem 6.22   The map $\sigma^+: \sum_k^+ \to \sum_k^+$ is weakly topologically mixing.

Proof. It suffices to prove the following: there exist $u, v \in \sum_k^+$ such that for any $s, t \in \sum_k^+$, there exists an increasing integer sequences ℓ_n such that

$$\lim_{n \to \infty} (\sigma^+ \times \sigma^+)^{\ell_n}(u, v) = (s, t),$$

i.e.,

$$\lim_{n \to \infty} (\sigma^+)^{\ell_n}(u) = s, \qquad \lim_{n \to \infty} (\sigma^+)^{\ell_n}(v) = t.$$

Let $n \geq 1$ and P_{nk^n} be any nk^n-block which was constructed in the proof of Theorem 6.19. There are totally $k^n!$ different nk^n-blocks. arranging them in any order, we have a $k^n!nk^n$-block, which is denoted by $Q_{k^n!nk^n}$.

Let

$$P_{k^n!nk^n} = \overbrace{P_{nk^n} \ldots P_{nk^n}}^{k^n! \text{ times}}.$$

(In total, $k^n! \, nk^n$-blocks P_{nk^n}.)

Define

$$u = (Q_{k!k} Q_{k^2!2k^2} \ldots Q_{k^n!nk^n} \ldots) \in \sum_k^+,$$
$$v = (P_{k!k} P_{k^2!2k^2} \ldots P_{k^n!nk^n} \ldots) \in \sum_k^+.$$

For any $n > 0$, we can see from the construction of u and v that there exists an $\ell_n > 0$ such that $(\sigma^+)^{\ell_n}(u)$ and s, $(\sigma^+)^{\ell_n}(v)$ and t, respectively, are identical up to the first $n + 1$ places. Then by (6.16), we have

$$d((\sigma^+)^{\ell_n}(u), s) \leq \frac{1}{2^n} \to 0, \qquad (n \to \infty),$$
$$d((\sigma^+)^{\ell_n}(v), t) \leq \frac{1}{2^n} \to 0, \qquad (n \to \infty).$$

Thus,

$$\lim_{n \to \infty} (\sigma^+)^{\ell_n}(u) = s, \qquad \lim_{n \to \infty} (\sigma^+)^{\ell_n}(v) = t. \tag{6.17}$$

□

Definition 6.23   Let f be continuous from a compact metric space X into itself. We say that f is topologically mixing if for any nonempty open sets $U, V \in X$, there exists a positive integer N such that

$$f^n(U) \cap V \neq \emptyset, \qquad \forall n > N.$$

□

Exercise 6.24   Verify the following implications [75]:

Topological mixing $\Rightarrow$ Weakly topological mixing $\Rightarrow$ Topological transitivity.

□

Theorem 6.25   The map $\sigma^+ \colon \Sigma_k^+ \to \Sigma_k^+$ is topologically mixing.

Proof.   Let $U, V \in \Sigma_k^+$ be any two nonempty open sets, and

$$s = (s_0 s_1 \ldots) \in U.$$

Then there exists an $\varepsilon_0 > 0$ such that

$$O_{\varepsilon_0}(s) = \left\{ t \in \sum_k^+ \;\middle|\; d(s, t) < \varepsilon_0 \right\} \subset U.$$

Let N be a sufficiently large integer such that $2^{-N} < \varepsilon_0$. Then by (6.16), we have

$$0[s_0 s_1 \ldots s_{N-1}] \equiv \left\{ t = (t_0 t_1 \ldots) \in \sum_k^+ \;\middle|\; t_i = s_i, \quad 0 \leq i \leq N - 1 \right\} \subset O_{\varepsilon_0}(s) \subset U.$$

It is easy to see that $(\sigma^+)^n(0[s_0 s_1 \ldots s_{N-1}]) = \Sigma_k^+$ when $n \geq N$. Therefore,

$$(\sigma^+)^n(U) \cap V \supset (\sigma^+)^n(0[s_0 s_1 \ldots s_{N-1}]) \cap V \supset \sum_k^+ \cap V = V \neq \emptyset,$$

if $n \geq N$.

□

Actually, by Exercise 6.24, we know that topologically mixing, implies weakly topologically mixing, and the latter implies topologically transitive. Thus, Theorem 6.25 implies Theorem 6.22, and Theorem 6.22 implies Theorem 6.19. Here we give the details of proofs of the theorems in order to help the readers get a deeper understanding in the symbolic dynamical system, in particular, how to construct a particular symbol sequence with a desired property.

Theorem 6.26   ([B.2])   Let $f: X \to X$ satisfy (6.7) that

(i) the set of all f's periodic points is dense,

(ii) f is topologically transitive,

(iii) f has sensitive dependence on initial data.

Then conditions (i) and (ii) imply condition (iii).

Proof. When X is a finite set, it must be a periodic orbit of f under conditions (i) and (ii). Thus, f obviously satisfies condition (iii). In the following, we assume that X is an infinite set.

We first claim that there exists a $\delta_0 > 0$ such that for every $x \in X$, there exists a periodic point p such that

$$d(\mathrm{orb}(p), x) \geq \frac{\delta_0}{2}.$$

In fact, take two different periodic points p_1 and p_2 which are on different periodic orbits, respectively. Let

$$\delta_0 = d(\mathrm{orb}(p_1), \mathrm{orb}(p_2)) \equiv \inf\{d(x, y) \mid x \in \mathrm{orb}(p_1), y \in \mathrm{orb}(p_2)\} > 0.$$

The triangle inequality

$$\delta_0 = d(\mathrm{orb}(p_1), \mathrm{orb}(p_2)) \leq d(\mathrm{orb}(p_1), x) + d(\mathrm{orb}(p_2), x),$$

implies either

$$d(\mathrm{orb}(p_1), x) \geq \frac{\delta_0}{2}$$

or

$$d(\mathrm{orb}(p_2), x) \geq \frac{\delta_0}{2}.$$

This proves our claim.

Next, we shall show that $\delta \triangleq \frac{\delta_0}{8}$ is a sensitivity constant for f. Let x be a given point and $0 < \varepsilon < \delta$. Then there exists a periodic point $p \in V(x, \varepsilon)$ by the denseness of periodic points, where $V(x, \varepsilon)$ is the ε-neighborhood of x. Denote by n the period of p.

From the claim above, there exists a periodic point q such that

$$d(\mathrm{orb}(q), x) \geq 4\delta.$$

Let

$$U = \bigcap_{i=1}^{n} f^{-1}(V(f^i(q)), \delta).$$

Then U is a nonempty open set containing q. It follows from the topological transitivity of f that there exist $y \in V(x, \varepsilon)$ and $k > 0$ such that $f^k(y) \in U$.

Let $j = [\frac{k}{n} + 1]$, where $[a]$ denotes the largest integer that is no larger than a. Then $1 \leq nj - k \leq n$. We have

$$f^{nj}(y) = f^{nj-k}(f^k(y)) \in f^{nj-k}(U) \subset V(f^{nj-k}(q), \delta).$$

Since $f^{nj}(p) = p$, we have

$$
\begin{aligned}
d(f^{nj}(p), f^{nj}(y)) &= d(p, f^{nj}(y)) \\
&\geq d(x, f^{nj-k}(q)) - d(f^{nj-k}(q), f^{nj}(y)) - d(p, x) \\
&> 4\delta - \delta - \delta = 2\delta.
\end{aligned}
$$

Again, by the triangle inequality

$$2\delta < d(f^{nj}(p), f^n j(y)) = d(f^{nj}(x), f^{nj}(y)) + d(f^{nj}(x), f^{nj}(p))$$

implies either

$$d(f^{nj}(x), f^{nj}(y)) > \delta,$$

or

$$d(f^{nj}(x), f^{nj}(p)) > \delta. \tag{6.18}$$

$\square$

Based on what Devaney [20] first gave in his book [20, Definition 8.5, p. 50] and Theorem 6.26, we now give the following definition of chaos.

Definition 6.27   Let (X, d) be a metric space and $f: X \to X$ be continuous. We say that f is *chaotic in the sense of Devaney* if

(i) f is topologically transitive; and

(ii) the set of all periodic points of f is dense. $\square$

Summarizing Theorems 6.18, 6.19 and 6.26, we see that the one-side shift map σ^+ is chaotic in the sense of Devaney.

Later in this section, we show that σ^+ is also chaotic in the sense of Li–Yorke. To to this end, we need some notations and lemmas. Without loss of generality, in the following, we assume that $k = 2$, i.e., 2-symbol dynamics.

We recall a famous result on chaos of interval maps. In 1975, Li and Yorke [49] obtained that for an interval map, period three implies chaos. More precisely, they proved the following.

Theorem 6.28   ([49]) Let $f: I \to I$ be continuous. Assume that there exists an $a \in I$ such that

$$f^3(a) \le a < f(a) < f^2(a), \quad \text{or} \quad f^3(a) \ge a > f(a) > f^2(a).$$

Then we have

(1) for every positive integer n, f has a periodic point with period n;

(2) there exists an uncountable set $C \subset I \backslash P(f)$ with the following properties:

   (i) $\limsup\limits_{n\to\infty} |f^n(x) - f^n(y)| > 0, \qquad \forall x, y \in C, x \ne y;$

   (ii) $\liminf\limits_{n\to\infty} |f^n(x) - f^n(y)| = 0, \qquad \forall x, y \in C;$

   (iii) $\limsup\limits_{n\to\infty} |f^n(x) - f^n(p)| > 0, \qquad \forall x \in C, \forall p \in P(f).$ □

The theorem of Li and Yorke motivates the following definition.

Definition 6.29   Let (X, d) be a metric space and $f: X \to X$ be continuous. We say that f is *chaotic in the sense of Li–Yorke* if

(1) for every positive integer n, f has a periodic point with period n;

(2) there exists an uncountable set $C \subset X \backslash P(f)$ with the following properties:

   (i) $\limsup\limits_{n\to\infty} d(f^n(x), f^n(y)) > 0, \qquad \forall x, y \in C, x \ne y;$

   (ii) $\liminf\limits_{n\to\infty} d(f^n(x), f^n(y)) = 0, \qquad \forall x, y \in C;$

   (iii) $\limsup\limits_{n\to\infty} d(f^n(x), f^n(p)) > 0, \qquad \forall x \in C, \forall p \in P(f).$ □

For every $s = (s_0 s_1 \ldots) \in \sum_2^+$, define

$$r(s, \ell) = \#\{i \mid s_i = 0, \quad i = 0, 1, \ldots, \ell\},$$

where $\#A$ denotes the cardinality of the set A.

Lemma 6.30   For every $0 < \eta < 1$, there exist an $s^\eta \in \sum_2^+$ and an integer ℓ, such that

$$\lim_{\ell \to \infty} \frac{r(s^\eta, \ell^2)}{\ell} = \eta.$$

Proof. Let $\ell_0 > 0$ be the smallest integer such that $[\ell_0 \eta] = 1$. Define

$$s^\eta = (s_0^\eta s_1^\eta \ldots s_\ell^\eta \ldots) \in \sum_2^+$$

as follows: For $0 \le i \le \ell_0^2$,

$$s_i^\eta = \begin{cases} 1, & 0 \le i < \ell_0^2, \\ 0, & i = \ell_0^2. \end{cases}$$

For $\ell_0^2 < i \le (\ell_0 + 1)^2$,

$$s_i^\eta = \begin{cases} 1, & \ell_0^2 < i < (\ell_0 + 1)^2, \\ 0, & \text{if } i = (\ell_0 + 1)^2 \text{ and } [(\ell_0 + 1)\eta] - [\ell_0 \eta] = 1, \\ 1, & \text{if } i = (\ell_0 + 1)^2 \text{ and } [(\ell_0 + 1)\eta] - [\ell_0 \eta] = 0. \end{cases}$$

Inductively, for $(\ell_0 + k)^2 < i \le (\ell_0 + k + 1)^2, k = 0, 1, \ldots,$

$$s_i^\eta = \begin{cases} 1, & (\ell_0 + k)^2 < i < (\ell_0 + k + 1)^2, \\ 0, & \text{if } i = (\ell_0 + k + 1)^2 \text{ and } [(\ell_0 + k + 1)\eta] - [(\ell_0 + k)\eta] = 1, \\ 1, & \text{if } i = (\ell_0 + k + 1)^2 \text{ and } [(\ell_0 + k + 1)\eta] - [(\ell_0 + k)\eta] = 0. \end{cases}$$

By the construction of s^η, it is easy to see that

$$r(s^\eta, \ell^2) = [\ell \eta], \qquad \forall \ell > 0.$$

Since $\ell \eta \le [\ell \eta] \le \ell \eta + 1$, we have

$$\lim_{\ell \to \infty} \frac{r(s^\eta, \ell^2)}{\ell} = \lim_{\ell \to \infty} \frac{[\ell \eta]}{\ell} = \eta. \tag{6.19}$$

$\square$

Lemma 6.31   Let s^η be defined as above. Then we have

(1) $s_i^\eta = 0$ if and only if there exists an $\ell > 0$ such that $i = \ell^2$ and $[\ell \eta] - [(\ell - 1)\eta] = 1$;

(2) for any $\ell \ge 1$, we have
$$s_{\ell^2 + j}^\eta = 1, \qquad 0 < j < 2\ell + 1;$$

(3) there exist infinite many integers $\ell > 0$ such that $s_{\ell^2}^\eta = 0$;

(4) let $0 < \eta < \theta < 1$. Then for any $N > 0$, there exists an $\ell > N$ such that $s_{\ell^2}^\theta \ne s_{\ell^2}^\eta$.

Proof. (1) and (2) follow from the construction of s^η.

For (3), suppose that there are only finite many integers $\ell > 0$ such that $s_{\ell^2}^\eta = 0$. Then it is easy to see that $r(s^\eta, \ell^2)$ is bounded by the construction of s^η. From Lemma 6.30, we have

$$\eta = \lim_{\ell \to \infty} \frac{r(s^\eta, \ell^2)}{\ell} = 0.$$

This is a contradiction.

For (4), suppose that it is not true. Then $r(s^\eta, \ell^2) - r(s^\theta, \ell^2)$ is bounded, which implies $\eta = \theta$ by Lemma 6.31, a contradiction. $\qquad\square$

Let (X, d) be a metric space and $f: X \to X$ be continuous. Recall that for $x \in X$, $y \in X$ is said to be a ω-*limit point of* x if there exists an increasing positive integer sequence n_i such that

$$\lim_{i \to \infty} f^{n_i}(x) = y.$$

Definition 6.32   Denote by $\omega(x, f)$ the set of all ω-limit points of x and call it the ω-limit set of x. $\qquad\square$

Lemma 6.33   Let (X, d) be a compact metric space and $f: X \to X$ be continuous. We have

(1) $\omega(x, f)$ is a nonempty closed set for every $x \in X$;

(2) for every $x \in X$;

$$f(\omega(x, f)) = \omega(x, f) = \omega(f^n(x), f), \qquad \forall n > 0.$$

Proof. Exercise. $\qquad\square$

Lemma 6.34   Let

$$C_0 \equiv \left\{ s^\eta \in \sum\nolimits_2^+ \ \middle|\ \eta \in (0, 1) \right\}, \qquad C = \bigcup_{i=0}^\infty (\sigma^+)^i(C_0),$$

where s^η is defined as in Lemma 6.30. Then we have

$$\omega(s, \sigma^+) = \{e, e_i \mid i = 0, 1, 2, \ldots\}, \qquad \forall s \in C, \tag{6.20}$$

where $e, e_i \in \sum_2^+$ are given by

$$e = (1, 1, 1, \ldots, 1, \ldots), \tag{6.21}$$
$$e_i = (1, 1, \ldots, 1, 0, 1, 1, \ldots).$$
$$\underset{\uparrow}{\text{the } i\text{-th bit}}$$

Proof. By Lemma 6.33, it suffices to prove (6.20) for $s \in C_0$. For $s^\eta \in C_0$, by Lemma 6.31, part (2), it is easy to see that

$$\lim_{\ell \to \infty} d((\sigma^+)^{\ell^2+1}(s^\eta), e) = 0.$$

Thus, $e \in \omega(s^\eta, \sigma^+)$.

By part (3) in Lemma 6.31, there exists an increasing sequence

$$\ell_1 < \ell_2 < \cdots < \ell_j < \cdots,$$

such the first bit in $(\sigma^+)^{\ell_j^2}(s^\eta)$ is 0. From parts (1) and (2) in Lemma 6.31, we have, for any $i > 0$, the first i bits are 1 and the $i + 1$-th is 0 in $(\sigma^+)^{\ell_j^2-i}(s^\eta)$ when j is larger enough. At the same time, there are at least $(\ell_j + 1)^2 - \ell_j^2 - 1 - (i + 1) = 2\ell_j - (i + 1)$ number of bits of 1 following the 0 bit. Thus,

$$\lim_{j \to \infty} d((\sigma^+)^{\ell_j^2-i}(s^\eta), e_i) = 0.$$

So we have $e_i \in \omega(s^\eta, \sigma^+)$, $\quad \forall i > 0$.

To complete our proof, it suffices to show that if $t \in \sum_2^+$ has at least two bits of 0, say $t_h = t_m = 0$, $h < m$, then

$$t \notin \omega(s^\eta, \sigma^+), \qquad \forall s^\eta \in C_0.$$

In fact, suppose that there exists an $\eta \in (0, 1)$ such that $t \in \omega(s^\eta, \sigma^+)$. Then, by definition, there exists an increasing sequence $\{\ell_j\}$ such that

$$\lim_{j \to \infty} (\sigma^+)^{\ell_j}(s^\eta) = t.$$

This implies that the $(h + 1)$-th and $(m + 1)$-th bits in $(\sigma^+)^{\ell_j}(s^\eta)$ are both 0 when j is large enough. This contradicts the fact that in $(\sigma^+)^{\ell_j}(s^\eta)$, there are infinite many 1 bits between any two 0 bits. $\qquad \square$

Theorem 6.35 Let C be defined as Lemma 6.30. We have

(1) for every positive integer n, σ^+ has a periodic point with period n;

(2) the set $C \subset \sum_2^+ \backslash P(\sigma^+)$ is uncountable, $\sigma^+(C) \subset C$ and has the following properties:

    (i) $\limsup\limits_{n \to \infty} d((\sigma^+)^n(s), (\sigma^+)^n(t)) > 0$, $\qquad \forall s, t \in C, s \neq t$;

    (ii) $\liminf\limits_{n \to \infty} d((\sigma^+)^n(s), (\sigma^+)^n(t)) = 0$, $\qquad \forall s, t \in C$;

    (iii) $\liminf\limits_{n \to \infty} d((\sigma^+)^n(s), (\sigma^+)^n(p)) > 0$, $\qquad \forall s \in C, \forall p \in P(\sigma^+)\backslash\{e\}$, where e is given in (6.21).

Thus, σ^+ is chaotic in the sense of Li–Yorke. $\qquad\qquad\qquad\qquad\qquad\qquad\qquad\square$

Comparing part (iii) in this theorem with Definition 6.29, we can see that σ^+ has chaotic dynamics stronger than Li–Yorke's chaos. Furthermore, the chaos set C for σ^+ can be taken to be invariant.

Proof of Theorem *6.35.* Here (1) follows from part (1) in Theorem 6.18.

For (2), first of all, since $s^\eta \neq s^\theta$ if $\eta \neq \theta$ from part (4) in Lemma 6.31, there is a one-to-one correspondence and between the open interval $(0, 1)$ and the set C_0. Thus, C_0 is uncountable, so is C. Obviously, $\sigma^+(C) \subset C$. From the construction of s^η, it is easy to see that for any $\ell > 0$, s^η and $(\sigma^+)^\ell(s^\eta)$ are not the periodic points of σ^+. So $C \subset \sum_2^+ \backslash \mathrm{Per}(\sigma^+)$.

Next, we prove parts (i)–(iii) in (2).

For (i), let $s, t \in C$, $s \neq t$, then there exist $s^\eta, s^\theta \in C_0$ with $0 < \eta \leq \theta < 1$ and two nonnegative integers h, m such that

$$s = (\sigma^+)^h(s^\eta), \qquad t = (\sigma^+)^h(s^\theta).$$

Assume that $0 \leq h \leq m$ ($h < m$ if $\eta = \theta$).

When $h = m$, we have $0 < \eta < \theta < 1$, and by part (4) in Lemma 6.31, there exist $\ell_1 < \ell_2 < \cdots < \ell_j < \cdots$ such that $s_{\ell_j^2}^\eta \neq s_{\ell_j^2}^\theta$, for $j = 1, 2, \ldots$. Thus,

$$d((\sigma^+)^{\ell_j^2 - h}((\sigma^+)^h(s^\eta)), (\sigma^+)^{\ell_j^2 - h}((\sigma^+)^h(s^\theta))) = d((\sigma^+)^{\ell_j^2}(s^\eta), (\sigma^+)^{\ell_j^2}(s^\theta)) \geq 1.$$

Therefore,

$$\limsup_{n \to \infty} d((\sigma^+)^n(s), (\sigma^+)^n(t)) \geq 1.$$

We now consider the case that $h < m$. From part (3) in Lemma 6.31, it follows that there exist $\ell_1 < \ell_2 < \cdots < \ell_j < \cdots$ such that $s_{\ell_j^2}^\eta = 0$. Thus, for any $j > 0$, the first bit in $(\sigma^+)^{\ell_j^2 - h}((\sigma^+)^h(s^\eta))$ is 0. On the other hand, there exists a j_0 such that for $j > j_0$, we have

$$\ell_j^2 < \ell_j^2 + m - h < (\ell_j + 1)^2 - 1.$$

Thus, if $j > j_0$, the $(\ell_j^2 + m - h + 1)$-th bit in s^θ must be 1 by part (1) in Lemma 6.31. This implies that

$$\limsup_{n \to \infty} d((\sigma^+)^n(s), (\sigma^+)^n(t))$$
$$\geq \limsup_{j \to \infty} d((\sigma^+)^{\ell_j^2 - h}((\sigma^+)^h(s^\eta)), (\sigma^+)^{\ell_j^2 - h}((\sigma^+)^m(s^\theta))) \geq 1.$$

So (i) is proved.

For (ii), let s, t be given as in (i). Let ℓ_0 be such that

$$\ell_0^2 \leq m - h + 1 \leq (\ell_0 + 1)^2.$$

For $j > j_0$, we consider

$$(\sigma^+)^{\ell^2 - h + 1}((\sigma^+)^h(s^\eta)) = (\sigma^+)^{\ell^2 + 1}(s^\eta),$$
$$(\sigma^+)^{\ell^2 - h + 1}((\sigma^+)^m(s^\theta)) = (\sigma^+)^{\ell^2 - h + m + 1}(s^\eta).$$

From part (2) in Lemma 6.31, the above two elements must agree with each other in the first $(\ell + 1)^2 - \ell^2 - (m - h + 1) = 2\ell - (m - h)$ bits. Thus,

$$\liminf_{n \to \infty} d((\sigma^+)^n(s), (\sigma^+)^n(t)) = 0.$$

Finally, for (iii), let $s = (\sigma^+)^h(s^\eta) \in C$ for some $s^\eta \in C_0$. Assume, on the contrary, that there exists a $p \in \text{Per}(\sigma^+)$ such that

$$\liminf_{n \to \infty} d((\sigma^+)^n((\sigma^+)^h(s^\eta)), (\sigma^+)^n(p)) = 0.$$

It follows from the periodic property of p that $p \in \omega(s^\eta, \sigma^+)$. But by Lemma 6.34, we know that e_i is not a periodic point of σ^+ for any $i > 0$. Thus, $p = e$. This completes the proof of (iii)   □

Exercise 6.36   Define $d_1(\cdot, \cdot)$ on $\sum_k^+$ by

$$d_1(s, t) = \max_n \left\{ \frac{1}{n + 1} \mid s_n \neq t_n \right\},$$

for $s = (s_0 s_1 \ldots), t = (t_0 t_1 \ldots \in \sum_k^+)$. Show that d_1 is a metric and is equivalent to d.   □

Exercise 6.37   Show that the shift map $\sigma^+ : \sum_k^+ \to \sum_k^+$ is k to one.   □

Exercise 6.38   Let $A = (a_{ij})$ be a k by k matrix with a_{ij} equal to 0 or 1, and there is at least an entry 1 in each row and each column. Define a subset Λ_A of $\sum_K^+$ by

$$\Lambda_A = \left\{ s = (s_0 s_1 \ldots) \in \sum_k^+ \mid a_{s_i s_{i+1}} = 1, \forall i \geq 0 \right\}.$$

Show that Λ_A is a nonempty closed invariant set of σ^+. That is, Λ_A is nonempty, closed and

$$\sigma^+(\Lambda_A) \subset \Lambda_A.$$

Such an matrix A is called a *transition matrix*.   □

Exercise 6.39   Consider a dynamical system (Λ_A, σ_A^+), where σ_A^+ is the restriction of σ^+ to Λ_A and A is defined as in Exercise 6.38. Prove that σ_A^+ is topologically transitive if and only if A is irreducible.   □

Exercise 6.40   Show that the following conditions are equivalent:

(i) σ_A^+ is topologically mixing;

(ii) σ_A^+ is topological weakly mixing;

(iii) A is aperiodic. (A transition matrix A is called *aperiodic* if there exists a positive integer n such that $A^n \gg 0$. That is, any entry of A^n is larger than 0.)   □

6.5   TOPOLOGICAL CONJUGACY AND SEMICONJUGACY

The concept of conjugacy arises in many subjects of mathematics. It also is an important concept in dynamical systems. All topological dynamical systems have different behaviors classifiable into various types of equivalence relations by topological conjugacy. Each class has the same topological dynamics.

Let X and Y be two metric spaces and h be a map from X to Y. Recall that h is called onto if for each $y \in Y$ there is some $x \in X$ with $h(x) = y$, h is called one-to-one if $h(x_1) \neq h(x_2)$ whenever $x_1 \neq x_2$. Also, recall that a map h from X to Y is called a *homeomorphism* provided that h is continuous, one-to-one and onto and its inverse h^{-1} from Y to X is also continuous.

Recalling Definition 5.8, let f mapping from X to X and g mapping from Y to Y be continuous. We say that f and g are *topologically conjugate*, or just conjugate (denoted by $f \simeq g$), if there exists a homeomorphism h from X to Y such that the following diagram commutes:

$$
\begin{array}{ccc}
X & \xrightarrow{f} & X \\
h\downarrow & & \downarrow h \\
Y & \xrightarrow{g} & Y
\end{array}
$$

i.e.,

$$hf = gh.$$

Such an h is called a *topological conjugacy*, from f to g. If h is merely a continuous map from X onto Y, then h is called a *semi-conjugacy*, and we say that f is semi-conjugate to g.

We give some properties of topological conjugacy below. We first note that topological conjugacy is an equivalence relation. That is,

(i) $f \simeq f$: The identity $h = id$ from X to X is a topological conjugacy from f to itself;

(ii) if $f \simeq g$, then $g \simeq f$. In fact, if h is a topological conjugacy from f to g, then its inverse h^{-1} is a topological conjugacy from g to f;

(iii) if $f_1 \simeq f_2$ and $f_2 \simeq f_3$, where f_3 is a continuous map from a metric space Z to itself, then $f_1 \simeq f_3$. In fact, let h_1 and h_2 be the topological conjugacies from f_1 to f_2 and from f_2 to f_3, respectively. Then $h = h_2 h_1$ is a topological conjugacy from f_1 to f_3.

Next, from $hf = gh$, it follows that

$$hf^2 = (hf)f = (gh)f = g(hf) = g(gh) = g^2h,$$

thus we see f^2 and g^2 are conjugated by h. Continuing by induction:

$$hf^n = (hf^{n-1})f = (g^{n-1}h)f$$
$$= g^{n-1}(hf) = g^{n-1}(gh) = g^nh,$$

thus f^n and g^n are conjugated by h for any $n > 0$.

Let $x_0 \in X$ and $x_n = f^n(x_0)$ be the point on the orbit generated by x_0 under f. Let $y_0 = h(x_0)$. Then

$$y_n = g^n(y_0)$$
$$= hf^nh^{-1}(y_0)$$
$$= hf^n(h^{-1}(y_0))$$
$$= hf^n(x_0) = h(x_n).$$

Thus, the orbit of x_0 under f is mapped to the orbit of $h(x_0)$ under g, i.e.,

$$h(\text{orb}(x, f)) = \text{orb}(h(x), g).$$

Lemma 6.41   Let f and g be conjugate by h. Then for every $x \in X$, we have

$$h(\omega(x, f)) = \omega(h(x), g).$$

Proof. Let $x_0 \in \omega(x, f)$. Then by the definition, there exists an increasing integer sequence $n_1 < n_2 < \cdots$ such that

$$\lim_{i \to \infty} f^{n_i}(x) = x_0.$$

Thus,

$$\lim_{i \to \infty} g^{n_i}(h(x)) = \lim_{i \to \infty} hf^{n_i}(x) = h(x_0),$$

and $h(x_0) \in \omega(h(x), g)$. So we have

$$h(\omega(x, f)) \subset \omega(h(x), g). \tag{6.22}$$

From (6.22), we have

$$\omega(x, f) \subset h^{-1}(\omega(h(x), g)). \tag{6.23}$$

On the other hand, since f and g are conjugated by h, g and f are conjugate by h^{-1}. It follows from (6.23) that

$$\omega(y, g) \subset h(\omega(h^{-1}(y), f)), \qquad \forall y \in Y,$$

and

$$\omega(h(x), g) \subset h(\omega(x, f)), \qquad \forall x \in X,$$

by taking $x = h^{-1}(y)$. Thus,

$$h(\omega(x, f)) = \omega(h(x), g). \tag{6.24}$$

$\square$

Theorem 6.42   Let f and g be conjugate by h. Then

(i) $p \in X$ is a periodic point of f with period m if and only if $h(p) \in Y$ is a periodic point of g with period m;

(ii) $h(\mathrm{Per}(f)) = \mathrm{Per}(g)$;

(iii) $\mathrm{Per}(f)$ is dense in X if and only if $\mathrm{Per}(g)$ is dense in Y;

(iv) f is topologically transitive if and only if g is topologically transitive.

Proof. For (i), let $p \in X$ be a periodic point of f with period m, i.e.,

$$f^i(p) \neq p, \qquad i = 1, 2, \ldots m - 1, \qquad f^m(p) = p.$$

Then

$$g^i(h(p)) = h(f^i(x)) \neq h(p), \qquad i = 1, 2, \ldots m - 1,$$

since h is one-to-one. And

$$g^m(h(p)) = h(f^m(p)) = h(p).$$

That is, $h(p)$ is a periodic point of g with period m.

The converse follows from the fact that g and f are conjugated by h^{-1}.

(ii) follows from (i).

For (iii), assume that $\mathrm{Per}(f)$ is dense in X. Let V be a nonempty open set in Y. Then $h^{-1}(V)$ is a nonempty open set in X. By the denseness of $\mathrm{Per}(f)$, there exists a $p \in \mathrm{Per}(f)$ such that $p \in h^{-1}(V)$. So $h(p) \in \mathrm{Per}(g)$ and $h(p) \in V$. That is, $\mathrm{Per}(g)$ is dense in Y. The converse follows from the fact that g and f are conjugated by h^{-1}.

For (iv), assume that f is topologically transitive, i.e., there exists an $x_0 \in X$ such that $\mathrm{orb}(x_0, f)$ is dense in X. Let V be any nonempty open set in Y. Then $h^{-1}(V)$ is a nonempty open set in X. Thus, there exists an n_0 such that

$$f^{n_0}(x_0) \in h^{-1}(V),$$

and

$$g^{n_0}(h(x_0)) = h(f^{n_0}(x_0)) \in V.$$

Thus, the orbit $\mathrm{orb}(h(x_0), g)$ is dense in Y.

The converse follows from the same way by noticing that g and f are conjugated by h^{-1}.   □

From Theorems 6.26 and 6.42, we have

Corollary 6.43   Let f and g be conjugate. Then f is chaotic in the sense of Devaney if and only if g is.

□

Theorem 6.44   Let f and g be conjugated by h. Then f is chaotic in the sense of Li–Yorke if and only if g is.

Proof. Assume that f satisfies Li–Yorke's chaos. From (i) in Theorem 6.42, it suffices to prove that g has property (2) in Definition 6.29.

To this end, assume that $C \subset X \backslash \mathrm{Per}(f)$ is an uncountable set with property (2) in Definition 6.29. We claim that $h(C) \subset Y \backslash \mathrm{Per}(g)$ is an uncountable set with property (2) with respect to g.

First, since C is uncountable and h is one-to-one, $h(C)$ is uncountable. And by (ii) in Theorem 6.42, $h(C) \subset Y \backslash \mathrm{Per}(g)$.

Next, for any $h(x), h(y) \in h(C), x, y \in C$ with $h(x) \neq h(y)$, we have $x \neq y$, since h is one-to-one. Thus,

$$\limsup_{n\to\infty} d_X(f^n(x), f^n(y)) > 0,$$

where d_X is the distance function on X. Since h is a homeomorphism, we have

$$\limsup_{n\to\infty} d_Y(g^n(h(x)), g^n(h(y))) = \limsup_{n\to\infty} d_X(h(f^n(x)), h(f^n(y))) > 0.$$

By the same reasoning, for any $x, y \in C$, we have

$$\liminf_{n\to\infty} d_Y(g^n(h(x)), g^n(h(y))) = \liminf_{n\to\infty} d_X(h(f^n(x)), h(f^n(y))) = 0.$$

Finally, for any $h(x) \in h(C)$ and any $h(p) \in \mathrm{Per}(g) = h(\mathrm{Per}(f))$, we have

$$\limsup_{n\to\infty} d_Y(g^n(h(x)), g^n(h(p))) = \limsup_{n\to\infty} d_X(h(f^n(x)), h(f^n(p))) > 0.$$

So g satisfies (2) in Definition 6.29.   □

Exercise 6.45   Definition: Let (X, f) and (Y, g) be compact dynamical systems and

$$f(X) = X, \qquad g(Y) = Y.$$

We say that f and g are topologically *semi-conjugate* if the h in Definition 6.27 is only continuous and onto. In this case, f is said to be an extension of g, g is a factor of f, and h is said to be a topological semi-conjugacy.

Assume that $h\colon X \to Y$ is a topological semi-conjugacy from f to g. Prove that if f is topologically transitive (resp., weakly mixing or mixing), so is g.   □

Exercise 6.46   Let $h\colon X \to Y$ be a topological semi-conjugacy. Prove that there exists a subset $X_h \subset X$ with the following properties:

(i) X_h is closed;

(ii) X_h is invariant set of f;

(iii) $h(X_h) = Y$;

(iv) There is no proper subset in X_h with the above three conditions.

Such an X_h is called an h-minimal cover of Y.

Let $h\colon X \to Y$ be a topological semi-conjugacy and X_h be an h-minimal cover of Y. Prove that if g is topologically transitive, so is $f|_{X_h}$.   □

6.6   SHIFT INVARIANT SETS

From the preceding sections, the dynamics of symbolic systems is now rather well understood. Especially, we know that σ and σ^+ have complex dynamical behaviors, such as manifesting chaos in the sense of both Devaney and Li–Yorke. Also, we have learned that two topologically conjugate systems have the same topological properties. In the rest of this chapter, we consider a class of topological systems which are conjugate to either the symbolic dynamical system $\sum_k^+$ or $\sum_k$.

6.7   CONSTRUCTION OF SHIFT INVARIANT SETS

Definition 6.47   Let X be a metric space and f be continuous from X to X. Let $\Lambda \subset X$ be a closed invariant set, i.e., $f(\Lambda) \subset \Lambda$. If the subsystem

$$f|_\Lambda\colon \ \Lambda \to \Lambda$$

is topologically conjugate to σ^+ or σ, i.e., there exists a homeomorphism h from Λ to Σ_k^+ or Σ_k such that the following diagram commutes

$$
\begin{array}{ccc}
\Lambda & \xrightarrow{f} & \Lambda \\
h\downarrow & & \downarrow h \\
\Sigma_k^+ & \xrightarrow{\sigma^+} & \Sigma_k^+
\end{array}
\qquad hf|_\Lambda = \sigma^+ h,
$$

then Λ is called a *shift invariant set* of f of order k.

If the h above is only continuous and onto, we call that Λ is a *quasi-shift invariant set* of order k.   □

Thus, if f has a shift invariant set, then it has complex behavior. In particular, it has chaos in the sense of both Devaney and Li–Yorke. In the following, we give necessary and sufficient conditions for a topological system to have a shift invariant set with respect to the one-sided shift σ^+. In the next chapter, we shall find the conditions that f has a shift invariant set with respect to the two-sided shift σ.

We need the following lemma.

Lemma 6.48   Let X, Y be two sets, f be a map from X to Y and $A \subset X, B \subset Y$. We have

$$
f(A \cap f^{-1}(B)) = f(A) \cap B.
$$

Proof. Straightforward verification.   □

Example 6.49   We now give an interval map that has a shift invariant set with respect to the one-sided shift. Consider the interval map $f : \mathbb{R} \to \mathbb{R}$:

$$
f(x) = -3x^2 + \frac{4}{3}.
$$

Let $J = [-1, 1]$. Then $f^{-1}(J)$ is composed of two subintervals U_0 and U_1 in J, which has an empty intersection:

$$
f^{-1}(J) = U_0 \cup U_1,
$$
$$
U_0 = \left[-\frac{\sqrt{7}}{3}, -\frac{1}{3} \right], \quad U_1 = \left[\frac{1}{3}, \frac{\sqrt{7}}{3} \right], \quad U_0 \cap U_1 = \emptyset,
$$

and

$$
|U_0| < \frac{1}{2}|J| = 1, \qquad |U_1| < \frac{1}{2}|J| = 1.
$$

See Fig. 6.3. Here $|J|$ is the length of the interval J. Thus, we have

$$f(U_0) = f(U_1) = J \supset U_0 \cup U_1. \tag{6.25}$$

From (6.25), there are two subinterval U_{00}, U_{01} and U_{10}, U_{11} in U_0 and U_1, respectively, with empty intersection such that

$$f(U_{00}) = U_0, \qquad f(U_{01}) = U_1, \tag{6.26}$$
$$f(U_{10}) = U_0, \qquad f(U_{11}) = U_1, \tag{6.27}$$

and

$$|U_{ij}| < \frac{1}{2}|U_j| < \frac{1}{2}, \qquad i, j = 0, 1. \tag{6.28}$$

Continuing by induction, for any positive integer k, we define

$$U_{s_0 s_1 \ldots s_k} = U_{s_0} \cap f^{-1}(U_{s_1}) \cap \cdots \cap f^{-k}(U_{s_k}), \tag{6.29}$$

where $s_0, s_1, \ldots, s_k \in \{0, 1\}$.

We continue the study of f throughout the following, up to equation (6.33).

$\square$

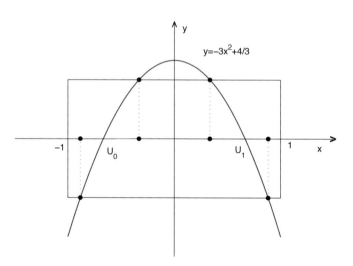

Figure 6.3: The map f and sets U_0, U_1 for Example 6.49.

Lemma 6.50 Let $U_{s_0 s_1 \ldots s_k}$ be defined as in (6.26). We have

(i) $f(U_{s_0 s_1 \ldots s_k}) = U_{s_1 \ldots s_k}$;

(ii) $|U_{s_0 s_1 \ldots s_k}| < \frac{1}{2^k}$.

Proof. Property (i) is true for $k = 1$ by (6.26)–(6.28).

Now we prove (i) and (ii) for the general case $k > 1$. Using Lemma 6.48 and $f(U_{s_0}) = J$, we have

$$f(U_{s_0 s_1 \ldots s_k}) = f(U_{s_0} \cap f^{-1}(U_{s_1 \ldots s_k}))$$
$$= f(U_{s_0}) \cap U_{s_1 \ldots s_k}$$
$$= U_{s_1 \ldots s_k}.$$

So (i) is verified.

For (ii), since f expands with an amplification factor greater than 2, we have

$$|U_{s_1 \ldots s_k}| = |f(U_{s_0 s_1 \ldots s_k})| > 2|U_{s_0 s_1 \ldots s_k}|.$$

Thus,

$$|U_{s_0 s_1 \ldots s_k}| < \frac{1}{2}|U_{s_1 \ldots s_k}| < \frac{1}{2} \cdot \frac{1}{2^{k-1}} = \frac{1}{2^k}. \tag{6.30}$$

$\square$

For $s = (s_0, s_1, s_2, \ldots) \in \Sigma_2^+$, define

$$U(s) = \bigcap_{j=0}^{\infty} f^{-j}(U_{s_j}) = \bigcap_{k=0}^{\infty} U_{s_0 s_1 \ldots s_k}. \tag{6.31}$$

Lemma 6.51   The $U(s)$ defined by (6.31) satisfies

(i) $f(U(s)) = U(\sigma^+(s))$;

(ii) $\sharp(U(s)) = 1$, i.e., $U(s)$ is a singleton set.

Proof. (i) From $f(U_{s_0}) = J$, it follows that

$$f(U(s)) = f\left(\bigcap_{j=0}^{\infty} f^{-j}(U_{s_j})\right)$$
$$= f(U_{s_0}) \bigcap_{j=1}^{\infty} f^{1-j}(U_{s_j})$$
$$= \bigcap_{k=1}^{\infty} U_{s_1 \ldots s_k}$$
$$= U(\sigma^+(s)).$$

(ii) It follows directly from (ii) in Lemma 6.50.

$\square$

Let

$$\Lambda = \bigcap_{j=0}^{\infty} f^{-j}(J) = \bigcup_{s \in \Sigma_2^+} U(s). \tag{6.32}$$

Theorem 6.52   The set Λ defined by (6.32) is a compact invariant set of f, and there exists a topological conjugacy h from σ^+ to $f|_\Lambda$. Thus, Λ is a compact shift invariant set of f.

Proof. It is easy to see from (6.32) and the compactness of Σ_2^+ that Λ is a compact invariant set of f.

We now construct a topological conjugacy from σ^+ to $f|_\Lambda$. Since, for every $s \in \Sigma_2^+$, the set $U(s)$ is a singleton by Lemma 6.51 part (ii), we may identify $U(s)$ as the singleton itself. Define $h: \Sigma_2^+ \to \Lambda$ by

$$h(s) = U(s).$$

It suffices to prove that h is a conjugacy from σ^+ to $f|_\Lambda$.

We first claim that h is a homeomorphism. In fact, for $s, t \in \Sigma_2^+$ with

$$d(s, t) < \frac{1}{2^n},$$

we have that s and t must agree with each other in the first n bits by (6.16). Thus,

$$h(s), h(t) \in U_{s_0 s_1 \ldots s_n} = U_{t_0 t_1 \ldots t_n},$$

and

$$|h(s) - h(t)| < \frac{1}{2^n}.$$

So h is continuous.

Let $s, t \in \Sigma_2^+, s \neq t$. Then there exists a $k > 0$ such that $s_k \neq t_k$. On the other hand, by the definition of $U(s)$, we have

$$h(s) \in U_{s_0 s_1 \ldots s_k}, \qquad h(t) \in U_{t_0 t_1 \ldots t_k},$$

and

$$f^k(h(s)) \in U_{s_k}, \qquad f^k(h(t)) \in U_{t_k}.$$

But

$$U_{s_k} \cap U_{t_k} = \emptyset.$$

Therefore,

$$f^k(h(s)) \neq f^k(h(t)),$$

and so

$$h(s) \neq h(t).$$

That is, h is one-to-one.

It is clear from the definition that h is onto Λ.

Since $\sum_2^+$ is compact space, Λ is Hausdorff and $h\colon \sum_2^+ \to \Lambda$ is one-to-one, onto and continuous, we have that h is a homeomorphism. This proves our claim.

Finally, it follows from (i) in Lemma 6.51 that

$$f|_\Lambda h = h\sigma^+. \tag{6.33}$$

That means that h is a topological conjugacy from σ^+ to $f|_\Lambda$.   □

From the above example, we see that condition (6.25) is the key in proving the existence of shift invariant set. In general, we have the following.

Theorem 6.53   Let X be a compact metric space and f be continuous from X to X. Then f has a shift invariant set of order k with respect to the one-sided shift σ^+ if and only if there exist compact subsets $A_0, A_1, \ldots, A_{k-1} \subset X$ which are pairwise disjoint such that

(i) $f(A_i) \supset \bigcup_{j=0}^{k-1} A_j$, $i = 0, 1, \ldots, k - 1$;

(ii) $\sharp\left(\bigcap_{j=0}^{\infty} f^{-j}(A_{s_j})\right) \leq 1$, $\forall (s_0 s_1 \ldots) \in \sum_k^+$.

Proof. We follow Zhang [74]. First, prove the necessity condition. Let Λ be a shift invariant set of f of order k, i.e., there exists a homeomorphism $h\colon \Lambda \to \sum_k^+$ such that

$$hf|_\Lambda = \sigma^+ h.$$

Denote

$$B_i = \left\{ s = (s_0 s_1 \ldots) \in \sum_k^+ \;\middle|\; s_0 = i \right\},$$
$$A_i = h^{-1}(B_i),   i = 0, 1, \ldots, k - 1.$$

It is easy to see that $A_0, A_1, \ldots, A_{k-1}$ are pairwise disjoint, closed subsets of Λ satisfying

(a)
$$f(A_i) = fh^{-1}(B_i)$$
$$= h^{-1}\sigma^+(B_i)$$
$$= h^{-1}\left(\sum_k^+\right)$$
$$= \Lambda \supset \bigcup_{j=0}^{k-1} A_j,   i = 0, 1, \ldots, k - 1.$$

(b) $$\bigcap_{j=0}^{\infty} f^{-j}(A_{s_j}) = \bigcap_{j=0}^{\infty} f^{-j} h^{-1}(B_{s_j})$$

$$= \bigcap_{j=0}^{\infty} h^{-1} (\sigma^+)^{-j}(B_{s_j})$$

$$= h^{-1} \bigcap_{j=0}^{\infty} (\sigma^+)^{-j}(B_{s_j})$$

$$= h^{-1}((s_0 s_1 \ldots)).$$

Since h is a homeomorphism, $h^{-1}((s_0 s_1 \ldots))$ is a singleton. Thus, we have

$$\sharp \left(\bigcap_{j=0}^{\infty} f^{-j}(A_{s_j}) \right) = 1,$$

and (i) and (ii) in the theorem hold.

Next, we show sufficiency. We first claim that for every $\ell > 0$ and any block $(s_0 s_1 \ldots s_\ell)$ in which every entry takes value in $\{0, 1, \ldots, k-1\}$, we have

$$f(A_{s_0 s_1 \ldots s_\ell}) = A_{s_1 \ldots s_\ell}, \tag{6.34}$$
$$f^\ell(A_{s_0 s_1 \ldots s_\ell}) = A_{s_\ell}, \tag{6.35}$$

where

$$A_{s_0 s_1 \ldots s_\ell} = \bigcap_{j=0}^{\ell} f^{-j}(A_{i_j}).$$

In fact, the proof of (6.34) is the same as that of Lemma 6.50. For (6.35), it is true for $\ell = 1$, since

$$f(A_{s_0} \cap f^{-1}(A_{s_1})) = f(A_{s_0}) \cap A_{s_1} = A_{s_1},$$

by Lemma 6.48 and Assumption (i).

Assume that it is also true for ℓ. Again, from Lemma 6.48, we have

$$
\begin{aligned}
f^{\ell+1}(A_{s_0 s_1 \ldots s_{\ell+1}}) &= f^{\ell+1}\left(\bigcap_{j=0}^{\ell+1} f^{-j}(A_{s_j})\right) \\
&= f \circ f^{\ell}\left(\bigcap_{j=0}^{\ell} f^{-j}(A_{s_j}) \cap f^{-\ell} \circ f^{-1}(A_{s_{\ell+1}})\right) \\
&= f\left(f^{\ell}\left(\bigcap_{j=0}^{\ell} f^{-j}(A_{s_j})\right) \cap f^{-1}(A_{s_{\ell+1}})\right) \\
&= f\left(A_{s_\ell} \cap A_{s_{\ell+1}}\right) \\
&= A_{s_{\ell+1}}.
\end{aligned}
$$

We establish that (6.35) is true for every $\ell > 0$ by induction. This proves our claim. From the compactness of X and (6.35), we have

$$
\bigcap_{j=0}^{\infty} f^{-j}(A_{s_j}) \neq \emptyset, \qquad \forall (s_0 s_1 \ldots) \in \sum_{k}^{+},
$$

by the nonempty intersection property of closed sets. Thus, assumption (ii) implies that for every $s = (s_0 s_1 \ldots) \in \Sigma_k^{+}$

$$
U(s) \equiv \bigcap_{j=0}^{\infty} f^{-j}(A_{s_j})
$$

is a singleton set. As before (in the proof of Theorem 6.52), we identify the set as the singleton contained therein.

Similar to part (i) in Lemma 6.51, we have

$$
f(U(s)) = U(\sigma^{+}(s)). \tag{6.36}
$$

Next, denote

$$
C = A_0 \cup \cdots \cup A_{k-1},
$$

and

$$
\Lambda = \bigcap_{j=0}^{\infty} f^{-j}(C) = \bigcup_{s \in \Sigma_k^{+}} U(s).
$$

Since C is compact and so is $f^{-j}(C)$ by the continuity of f, Λ is compact by Tychonov's Theorem [26]. And

$$
f(\Lambda) \subset \Lambda.
$$

Define $h \colon \Sigma_k^+ \to \Lambda$ by

$$h(s) = U(s).$$

Similar to the proof of Theorem 6.52, we can show that h is one-to-one and onto Λ. To show that h is a homeomorphism, it suffices to prove that either h or h^{-1} is continuous. Here we prove the latter for the sake of self-containedness, since we have already proved the continuity of h in Theorem 6.52.

For any $x \in \Lambda$, let

$$h^{-1}(x) = s = (s_0 s_1 \ldots).$$

For any $\varepsilon > 0$, by (6.16), there exists an $n > 0$ such that

$$[s_0 s_1 \ldots s_{n-1}]_0 \equiv \left\{ t = (t_0 t_1 \ldots) \in \sum_k^+ \;\middle|\; t_0 = s_0, \ldots, t_{n-1} = s_{n-1} \right\}$$

$$\subset \left\{ t \in \sum_k^+ \;\middle|\; d(s, t) < \varepsilon \right\}.$$

Since f is continuous, and since $A_0, \ldots, A_{k-1}$ has empty pairwise intersections and

$$f^j(x) \in A_{s_j}, \qquad j = 0, 1, \ldots, n-1,$$

there exists a $\delta > 0$ such that when $y \in O_\delta(x) \cap \Lambda$, we have

$$f^j(y) \in A_{s_j}, \qquad j = 0, 1, \ldots, n-1.$$

By noting that $y \in \Lambda$, we have

$$y = A_{s_0} \cap f^{-1}(A_{s_1}) \cap \cdots \cap f^{-(n-1)}(A_{s_{n-1}}) \bigcap_{j=n}^{\infty} f^{-j}(A_{i_j}),$$

for some $(i_n i_{n+1} \ldots) \in \Sigma_k^+$. Thus,

$$h^{-1}(y) \in [s_0 \ldots s_{n-1}]_0,$$

by the definition of h^{-1}. This prove the continuity of h^{-1}.

Finally, it follows from (6.36) that

$$f|_\Lambda h = h\sigma^+.$$

Thus, h is a topological conjugacy from f to σ^+. □

From the proof of Theorem 6.53, it is easy to see the following.

Corollary 6.54 Let X be a compact metric space and f be continuous from X to X. Then f has a quasi-shift invariant set of order k with respect to the one-sided shift if and only if condition (i) in Theorem 6.53 holds. □

6.8   SNAP-BACK REPELLER AS A SHIFT INVARIANT SET

In this section, we present a dynamical system on $\mathbb{R}^N$ that has a shift invariant set.

Let $x \in \mathbb{R}^N$ and f be a differentiable map from an open set $\mathcal{O} \subseteq \mathbb{R}^N$ into $\mathbb{R}^N$. Thus, $f = (f_1, f_2, \ldots, f_N)$, where $f_1, f_2, \ldots, f_N$ are the components of f. We denote

$$Df(x) = \begin{bmatrix} \dfrac{\partial f_1}{\partial x_1} & \dfrac{\partial f_2}{\partial x_2} & \cdots & \dfrac{\partial f_N}{\partial x_1} \\ \dfrac{\partial f_1}{\partial x_2} & \dfrac{\partial f_2}{\partial x_2} & \cdots & \dfrac{\partial f_N}{\partial x_2} \\ \vdots & & & \\ \dfrac{\partial f_1}{\partial x_N} & \dfrac{\partial f_2}{\partial x_N} & \cdots & \dfrac{\partial f_N}{\partial x_n} \end{bmatrix}(x).$$

Definition 6.55   Let $f \colon \mathbb{R}^N \longrightarrow \mathbb{R}^N$ be C^1. Let p be a fixed point of f. We say that p is a *repelling fixed point* if all of the eigenvalues of $Df(p)$ have absolute value larger than 1. A repelling fixed point p is called a *snap-back repeller* if, for any neighborhood V of p, there exist $q \in V, q \neq p$ and an integer m, such that $f^m(q) = p$ and $\det Df^m(q) \neq 0$. Such a point q is called a snap-back point.   □

The definition of a snap-back repeller was first introduced by Marotto [51] where he showed that it implies chaos. Here, we show that there exists a positive integer r such that f^r has a shift invariant set with respect to the one-sided shift if f has a snap-back repeller.

Lemma 6.56   Let A be an $N \times N$ constant matrix such that all of its eigenvalues have absolute values larger than μ. See also Marotto [52]. Then there exists a norm $|\cdot|$ in $\mathbb{R}^N$ such that the associated operator- norm of A^{-1}, $\|A^{-1}\|$, satisfies $\|A^{-1}\| \leq \mu$. Consequently, A satisfies $|Ax| \geq \mu|x|$ for all $x \in \mathbb{R}^N$.   □

Proof. This is left as an exercise.   □

For $x \in \mathbb{R}^N$, let $\mathcal{N}(x)$ denote the set of all closed neighborhoods V of x, where

$$V = \{y \in \mathbb{R}^N \mid |y - x| \leq r\}, \qquad r > 0.$$

Lemma 6.57   Let $f \colon \mathbb{R}^N \longrightarrow \mathbb{R}^N$ be C^1. If f has a snap-back repeller p, then there exists a $U \in \mathcal{N}(p)$ such that

(i) $U \subset f(U)$;

(ii) $\bigcap_{k=0}^{\infty} f^{-k}(U) = \{p\}$;

(iii) let $q \in \text{int } U$ be the associated snap-back point. Then there exists a $V \in \mathcal{N}(q)$ such that $f^m \colon V \to f^m(V)$ is a diffeomorphism.

Proof. Since p is a snap-back repeller, all of the eigenvalues of $Df(p)$ have absolute values larger than 1. By Lemma 6.56, we can define a norm $|\cdot|$ on $\mathbb{R}^N$ such that

$$|Df(p)x| \geq \mu |x|, \qquad \forall x \in \mathbb{R}^N, \tag{6.37}$$

for some $\mu > 1$.

Fix μ' with $1 < \mu' < \mu$. By continuity, we can find a $U \in \mathcal{N}(p)$ such that

$$\|Df(x) - Df(p)\| \leq \mu - \mu', \qquad \forall x \in U. \tag{6.38}$$

Let $\omega = (f - Df(p))|_U$. By (6.38), we have

$$|\omega(x') - \omega(x)| \leq \left(\int_0^1 \|D\omega(x + t(x' - x))\| dt \right) |x' - x|$$
$$\leq (\mu - \mu')|x' - x|, \qquad \forall x', x \in U.$$

Thus, for any $x', x \in U$, we have

$$|f(x') - f(x)| = |(f(x') - Df(p)x') - (f(x) - Df(p)x) + Df(p)(x' - x)|$$
$$\geq |Df(p)(x' - x)| - |\omega(x') - \omega(x)|$$
$$\geq \mu |x' - x| - (\mu - \mu')|x' - x| = \mu'|x' - x|. \tag{6.39}$$

Therefore, f expands U and (i) holds.

For (ii), since $f^{-1}(U) \subset U$ by (i), we have

$$\cdots \subset f^{-k}(U) \subset f^{-(k-1)}(U) \subset \cdots \subset f^{-1}(U) \subset U.$$

Also, from (6.39), it follows that

$$|f^{-1}(x') - f^{-1}(x)| \leq \frac{1}{\mu'}|x' - x|, \qquad \forall x', x \in U.$$

Thus,

$$|f^{-k}(x') - f^{-k}(x)| \leq \frac{1}{\mu'}|f^{-(k-1)}(x') - f^{-(k-1)}(x)|$$
$$\leq \cdots\cdots$$
$$\leq \frac{1}{\mu'^k}|x' - x| \to 0, \qquad \text{as} \quad k \to \infty,$$

for all $x', x \in U$. We have (ii).

Property (iii) follows from $\det Df^m(p) \neq 0$ and the inverse function theorem. $\qquad \square$

Theorem 6.58   Let $f : \mathbb{R}^N \longrightarrow \mathbb{R}^N$ be C^1. If f has a snap-back repeller at p, then there exists a positive integer r such that f^r has a shift invariant set of order 2 with respect to the one-sided shift.

Proof. Let U and V be defined as in Lemma 6.57. By Theorem 6.53, it suffices to prove that there exist two nonempty compact sets A_1, $A_2 \subset U$ with empty intersection such that (i) and (ii) in Theorem 6.53 hold with f therein substituted by f^r.

Let $q \in U$ be a snap-back point. Since $q \neq p$, there exist two disjoint nonempty neighborhoods $V_0 \in \mathcal{N}(p)$ and $W_0 \in \mathcal{N}(q)$ which are contained in U and $U \cap V$, respectively. Let $W = f^{-m}(V_0) \cap W_0$. Since f^m is continuous and $W \neq \emptyset$ (in fact $q \in W$), it follows that $W \in \mathcal{N}(q)$. Let $V_1 = f^m(W)$. Then $V_1 \in \mathcal{N}(p)$ since f^m is a diffeomorphism from V to $f^m(V)$. By Lemma 6.48, $V_1 = f^m(f^{-m}(V_0) \cap W_0) = V_0 \cap f^m(W_0)$. So $V_1 \cap W = \emptyset$.

On the other hand, since f^{-1} is a strict contraction on U, there exists an integer $l \geq 0$ such that

$$\bigcap_{j=0}^{\ell} f^{-j}(U) \subset V_1.$$

Let $r = m + \ell$, $A_1 = \bigcap_{j=0}^{r-1} f^{-j}(U)$ and $A_2 = W$. So $A_1 \cap A_2 \neq \emptyset$. It remains to prove that they satisfy (i) and (ii) in Theorem 6.53.

By applying Lemma 6.48 r times, we have

$$f^r(A_1) = f(U) \supset U \supset A_1 \cup A_2,$$

$$f^r(A_2) = f^\ell(f^m(W)) = f^\ell(V_1) \supset f^\ell\left(\bigcap_{j=0}^{\ell} f^{-j}(U)\right) \supset U \supset A_1 \cup A_2.$$

Thus, (i) in Theorem 6.53 holds.

Condition (ii) in Theorem 6.53 follows from the fact that f^{-r} is a strict contraction and that

$$\bigcap_{j=0}^{\infty} f^{-rj}(A_{s_j}) \subset \bigcap_{j=0}^{\infty} f^{-rj}(U) = \{p\},$$

for any $(s_0 s_1 \ldots) \in \Sigma_2^+$.   □

Exercise 6.59   Let $f \in C^1(I)$. If $|f'(x)| \leq 1$, $\forall x \in I$, prove that f has no shift invariant sets.   □

Exercise 6.60   Let $f(x) = 4x(1 - x)$. Show that there exist two non-empty closed subintervals K_0 and K_1 of $[0, 1]$ with empty intersection such that

$$f^2(K_0) \cap f^2(K_1) \supset [0, 1].$$ □

Exercise 6.61   Let $f \in C^0(I)$. Prove that if g has a periodic point whose period is not a power of 2, then there exist two non-empty closed subintervals K_0 and K_1 of I with empty intersection and a positive integer m such that

$$f^m(K_0) \cap f^m(K_1) \supset K_0 \cup K_1.$$

□

Exercise 6.62   Prove Corollary 6.54.

□

Exercise 6.63   Construct a piecewise continuous linear map f on I such that f has a quasi-shift invariant set of order k, for any positive integer k.

□

NOTES FOR CHAPTER 6

Symbolic dynamics originated as a method and representation to study general dynamical systems. The original ideas may be traced back to the works of several mathematicians: J. Hadamard, M. Morse, E. Artin, P.J. Myrberg, P. Koebe, J. Nielsen, and G.A. Hedlund. But the first formal treatment was made by Morse and Hedlund [55] in 1938.

A salient feature of symbolic dynamics is that time is measured in discrete time, and the orbits are represented by a string of symbols. The role of the system dynamics or time evolution becomes the shift operator. Important concepts such as periodicity, denseness of orbits, topological mixing, topological transitivity, and sensitive dependence on initial data all have particularly elegant discussions or representations in symbolic dynamics.

C. Shannon used symbolic sequences and shifts of finite type in his seminal paper [61] to study information theory in 1948. Today, symbolic dynamics finds many applications in computer science such as data storage, transmission and manipulation, and other areas.

Our treatment in Section 6.4 is mostly based on Zhou [75], in particular, Lemmas 6.30–6.34 and Theorem 6.35. Theorems 6.52–6.53 comes from Zhang [73, 74]. Theorems 6.58 comes from [15].

CHAPTER 7

The Smale Horseshoe

The Smale horseshoe offers a model for pervasive high-dimensional nonlinear phenomena, as well as a powerful technique for proving chaos. Here in this chapter, we present the famous Smale horseshoe and show that it has a shift invariant set with respect to the two-sided shift. We first introduce the standard Smale horseshoe and then discuss the general case.

7.1 THE STANDARD SMALE HORSESHOE

Let $\varepsilon > 0$ be small. Consider the two square regions P and Q in $\mathbb{R}^2$:

$$P = (-1 - \varepsilon, 1 + \varepsilon) \times (-1 - \varepsilon, 1 + \varepsilon), \tag{7.1}$$

and

$$Q = [-1, 1] \times [-1, 1]. \tag{7.2}$$

The horseshoe map is described geometrically on P as the composition of two maps. One map is linear which expands in vertical direction by λ ($\lambda > 2$) and contracts in horizontal direction by $1/\lambda$. The other is a nonlinear smooth map which bends the image of the linear map into a horseshoe-shaped object and places the image on P. See Fig. 7.1.

By the above construction, we have defined a horseshoe map

$$\varphi \colon P \to \mathbb{R}^2,$$

which is a diffeomorphism from P to $\varphi(P)$. We now show that φ has a shift invariant set of order 2 with respect to a two-sided shift.

First, we observe that

$$V = \varphi(Q) \cap Q$$

is composed of two vertical strips V_0 and V_1 with empty intersection, i.e.,

$$V = V_0 \cup V_1.$$

The width of each strip is less than the width of Q:

$$\theta(V_0) < 1, \qquad \theta(V_1) < 1.$$

Here $\theta(V_i)$ denotes twice the width of the vertical strip $V_i, i = 0, 1$.

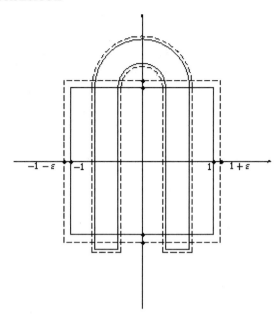

Figure 7.1: The standard Smale horseshoe. The square with dotted lines is P, while the square with the solid lines is Q.

Next, we observe that $U = \varphi^{-1}(V)$ is composed of two horizontal strips $U_0 = \varphi^{-1}(V_0)$ and $U_1 = \varphi^{-1}(V_1)$, i.e.,

$$U = U_0 \cup U_1, \quad \text{and} \quad U_0 \cap U_1 = \emptyset.$$

The thickness of each horizontal strip is less than half the thickness of Q:

$$\theta(U_0) < 1, \qquad \theta(U_1) < 1.$$

Here $\theta(U_i)$ denotes twice the thickness of the horizontal strip U_i. See Fig. 7.2.

In the following, denote

$$U_{ij} = \varphi^{-1}(V_i \cap U_j) = U_i \cap \varphi^{-1}(U_j), \qquad i, j = 0, 1;$$
$$V_{ij} = \varphi(U_i \cap V_j) = V_i \cap \varphi(V_j), \qquad i, j = 0, 1.$$

It is easy to see that U_{ij} is a horizontal strip contained in U_i. Its thickness is less than half the thickness of U_i:

$$\theta(U_{ij}) < \frac{1}{2}\theta(U_i) < \frac{1}{2}.$$

Likewise, V_{ij} is a vertical strip contained in V_i. Its width is less than half the width of V_i:

$$\theta(V_{ij}) < \frac{1}{2}\theta(V_i) < \frac{1}{2}.$$

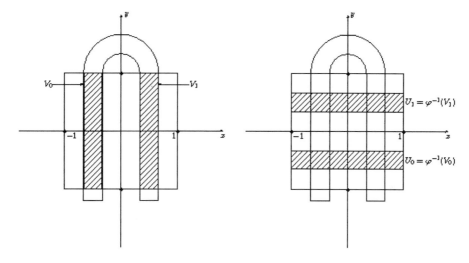

Figure 7.2: Illustrations of the sets V_0, V_1, and U_0, U_1.

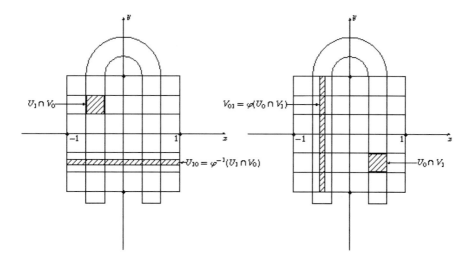

Figure 7.3: Illustration of the sets U_{10} and V_{01}.

See Fig. 7.3.

In general, for

$$s_{-k}, \ldots, s_{-1}, s_0, s_1, \ldots, s_k \in \{0, 1\} = S(2), \quad (\text{cf. } (6.11)),$$

and $k > 1$, we define $U_{s_0\ldots s_k}$ and $V_{s_{-k}\ldots s_{-2}s_{-1}}$ inductively as follows:

$$U_{s_0\ldots s_k} = \varphi^{-1}(V_{s_0} \cap U_{s_1\ldots s_k}) = U_{s_0} \cap \varphi^{-1}(U_{s_1\ldots s_k})$$
$$= U_{s_0} \cap \varphi^{-1}(U_{s_1}) \cap \cdots \cap \varphi^{-k}(U_{s_k});$$
$$V_{s_{-k}\ldots s_{-2}s_{-1}} = \varphi(U_{s_{-1}} \cap V_{s_{-k}\ldots s_{-2}}) = V_{s_{-1}} \cap \varphi(V_{s_{-k}\ldots s_{-2}})$$
$$= V_{s_{-1}} \cap \varphi(V_{s_{-2}}) \cap \cdots \cap \varphi^{k-1}(V_{s_{-k}}).$$

We can see that $U_{s_0\ldots s_k}$ is a horizontal strip contained in $U_{s_0\ldots s_{k-1}}$ and $V_{s_{-k}\ldots s_{-2}s_{-1}}$ is a vertical strip contained in $V_{s_{-k+1}\ldots s_{-2}s_{-1}}$.

Lemma 7.1   We have

(i) $\varphi(U_{s_0\ldots s_k}) = V_{s_0} \cap U_{s_1\ldots s_k}$, $\varphi(V_{s_{-k}\ldots s_{-2}s_{-1}}) \cap V_{s_0} = V_{s_{-k}\ldots s_{-1}s_0}$;

(ii) $\theta(U_{s_0\ldots s_k}) < \frac{1}{2}\theta(U_{s_1\ldots s_k}) < \frac{1}{2^k}, \theta(V_{s_{-k}\ldots s_{-2}s_{-1}}) < \frac{1}{2}\theta(V_{s_{-k}\ldots s_{-2}}) < \frac{1}{2^k}$.

Proof. (i) follows directly from the definition.
For (ii), by definition,

$$U_{s_0\ldots s_k} = \varphi^{-1}(V_{s_0} \cap U_{s_1\ldots s_k}).$$

Since φ^{-1} is a strict contraction in the vertical direction with a contracting rate less than $1/2$, we have

$$\theta(U_{s_0\ldots s_k}) < \frac{1}{2}\theta(U_{s_1\ldots s_k}) < \frac{1}{2^k}.$$

Similarly, we have

$$\theta(V_{s_{-k}\ldots s_{-2}s_{-1}}) < \frac{1}{2}\theta(V_{s_{-k}\ldots s_{-2}}) < \frac{1}{2^k}. \tag{7.3}$$

$\square$

For $s = (\ldots, s_{-2}, s_{-1}, s_0, s_1, s_2, \ldots) \in \Sigma_2$, we introduce

$$U(s) = \bigcap_{j=0}^{\infty} \varphi^{-j}(U_{s_j})$$
$$= \bigcap_{k=0}^{\infty} U_{s_0 s_1 \ldots s_k} \left(= \bigcap_{k=1}^{\infty} U_{s_0 s_1 \ldots s_k} \right),$$
$$V(s) = \bigcap_{j=1}^{\infty} \varphi^{j-1}(V_{s_{-j}})$$
$$= \bigcap_{k=1}^{\infty} V_{s_{-k}\ldots s_{-1}} \left(= \bigcap_{k=2}^{\infty} V_{s_{-k}\ldots s_{-1}} \right).$$

Lemma 7.2 We have

(i) $\varphi(V(s) \cap U(s)) = V(\sigma(s)) \cap U(\sigma(s))$;

(ii) $\sharp(V(s) \cap U(s)) = 1$.

Proof. (i) By part (i) in Lemma 7.1, it follows that

$$\varphi(U(s)) = \bigcap_{k=1}^{\infty} \varphi(U_{s_0 s_1 \ldots s_k})$$
$$= V_{s_0} \cap \bigcap_{k=1}^{\infty} U_{s_1 \ldots s_k}$$
$$= V_{s_0} \cap U(\sigma(s)),$$
$$\varphi(V(s)) \cap V_{s_0} = \bigcap_{k=1}^{\infty} \varphi(V_{s_{-k} \ldots s_{-1}}) \cap V_{s_0}$$
$$= \bigcap_{k=1}^{\infty} V_{s_{-k} \ldots s_{-1} s_0}$$
$$= V(\sigma(s)).$$

Therefore,

$$\varphi(V(s) \cap U(s)) = \varphi(V(s)) \cap \varphi(U(s))$$
$$= \varphi(V(s)) \cap V_{s_0} \cap U(\sigma(s))$$
$$= V(\sigma(s)) \cap U(\sigma(s)).$$

We have proved (i).

(ii) For any $k \in \mathbb{N}$, we have

$$V(s) \cap U(s) \subset V_{s_{-k} \ldots s_{-1}} \cap U_{s_0 \ldots s_k}.$$

But by part (ii) in Lemma 6.48, we have

$$\theta(V_{s_{-k} \ldots s_{-1}}) < \frac{1}{2^{k-1}}, \qquad \theta(U_{s_0 \ldots s_k}) < \frac{1}{2^k}.$$

So we have $\sharp(V(s) \cap U(s)) = 1$. $\qquad \square$

Denote

$$\Lambda = \bigcup_{s \in \Sigma_2} (V(s) \cap U(s)).$$

We identify the singleton set $V(s) \cap U(s)$ with the singleton itself contained therein and define a map $h\colon \sum_2 \to \Lambda$ by

$$h(s) = V(s) \cap U(s), \qquad \forall s \in \sum\nolimits_2.$$

Theorem 7.3 (Smale [64, 65]) The set Λ is a compact invariant set of φ, and $\varphi|_\Lambda$ is topologically conjugate to the two-sided shift $\sigma\colon \sum_2 \to \sum_2$ with conjugacy h. Therefore, the horseshoe map φ has a shift invariant set of order 2 with respect to the two-sided shift.

Proof. First, we claim that $h\colon \sum_2 \to \Lambda$ is continuous. In fact, if $s, t \in \sum_2$ satisfies

$$d(s, t) < \frac{1}{2^k},$$

Then $h(s)$ and $h(t)$ must belong to a rectangle which has its width less than $1/2^{k-1}$ and its thickness less than $1/2^k$:

$$h(s), h(s) \in V_{s_{-k}\dots s_{-1}} \cap U_{s_0 s_1 \dots s_k}.$$

So we have proved our claim.

Next, we prove that h is one-to-one. Let $s, t \in \sum_2$. If there exists a $k \in \mathbb{Z}, k \leq 0$, such that $s_k \neq t_k$, then $U_{s_k} \cap U_{t_k} = \emptyset$ and

$$\varphi^k(h(s)) \in U_{s_k}, \qquad \varphi^k(h(t)) \in U_{s_t}.$$

Thus, $h(s) \neq h(t)$.

If there exists an $\ell \in \mathbb{Z}, \ell > 0$, such that $s_{-\ell} \neq t_{-\ell}$, then $V_{s_{-\ell}} \cap V_{t_{-\ell}} = \emptyset$, and

$$\varphi^{-(\ell-1)}(h(s)) \in V_{s_{-\ell}}, \qquad \varphi^{-(\ell-1)}(h(t)) \in V_{t_{-\ell}}.$$

Thus, we also have $h(s) \neq h(t)$.

The proof of the remainders is the same as that of Theorem 6.52. $\square$

7.2 THE GENERAL HORSESHOE

Throughout, we use Q to denote the unit square $[-1, 1] \times [-1, 1]$ as in (7.1). The Smale horseshoe map discussed in Section 7.1 appears quite artificial and restrictive. In this section, we introduce the Conley–Moser condition which can also generate a "horseshoe" type shift invariant set, but is much more general.

Let $u\colon [-1, 1] \to [-1, 1]$ be a continuous function. We say that the curve $y = u(x)$ is a μ_h-horizontal curve if the function satisfies

$$|u(x_1) - u(x_2)| \leq \mu_h |x_1 - x_2|, \qquad \forall x_1, x_2 \in [-1, 1].$$

i.e., u is a Lipschitz function with Lipschitz constant μ_h. Similarly, for v: $[-1, 1]$ $\to [-1, 1]$, we say that $x = v(y)$ is a μ_v-vertical curve if

$$|v(y_1) - v(y_2)| \leq \mu_v |y_1 - y_2|, \qquad \forall y_1, y_2 \in [-1, 1].$$

Definition 7.4

(i) Let $y = u_1(x)$ and $y = u_2(x)$ be two non-intersecting μ_h-horizontal curves such that

$$-1 \leq u_1(x) < u_2(x) \leq 1.$$

We call

$$U = \{(x, y) \in \mathbb{R}^2 \mid -1 \leq x \leq 1, u_1(x) \leq y \leq u_2(x)\}$$

a μ_h-horizontal strip.

(ii) Let $x = v_1(y)$ and $x = v_2(y)$ be two non-intersecting μ_v-vertical curves such that

$$-1 \leq v_1(y) < v_2(y) \leq 1.$$

We call

$$V = \{(x, y) \in \mathbb{R}^2 \mid -1 \leq y \leq 1, v_1(y) \leq x \leq v_2(y)\}$$

a μ_v-vertical strip.

See Fig. 7.4 for an illustration.

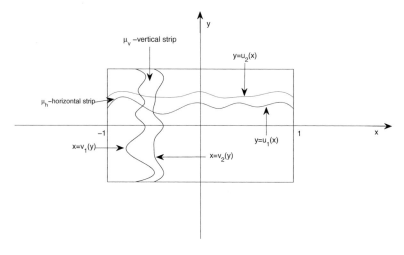

Figure 7.4: Illustration of a μ_v-vertical strip and a μ_h-horizontal strip.

We define the thickness of U and the width of V, respectively, by

$$\theta(U) = \max_{-1 \leq x \leq 1} \{u_2(x) - u_1(x)\}, \quad \theta(V) = \max_{-1 \leq y \leq 1} \{v_2(y) - v_1(y)\}.$$

Lemma 7.5   Let $U^1 \supset U^2 \supset \cdots \supset U^k \supset U^{k+1} \supset \cdots$ be a nested sequence of μ_h-horizontal strips such that

$$\lim_{k \to +\infty} \theta(U^k) = 0.$$

Then $U^\infty \equiv \bigcap_{k=1}^\infty U^k$ is a μ_h-horizontal curve.

   Similarly, for a nested sequence of μ_v-vertical strips $V^1 \supset V^2 \supset \cdots \supset V^k \supset V^{k+1} \supset \cdots$ satisfying $\lim_{k \to +\infty} \theta(U^k) = 0$, then $V^\infty \equiv \bigcap_{k=1}^\infty V^k$ is a μ_v-vertical curve.

Proof. Let

$$U^k = \{(x, y) \in \mathbb{R}^2 \mid -1 \leq x \leq 1, u_1^{(k)}(x) \leq y \leq u_2^{(k)}(x)\}.$$

By the given assumptions, for fixed $x \in [-1, 1]$, $\{u_1^{(k)}(x)\}$ is increasing and $\{u_2^{(k)}(x)\}$ is decreasing and $u_1^{(k)}(x) < u_2^{(k)}(x)$. There exists the squeezed limit

$$\lim_{k \to +\infty} u_1^{(k)}(x) = \lim_{k \to +\infty} u_2^{(k)}(x) \equiv u(x).$$

We thus obtain a function $u : [-1, 1] \to [-1, 1]$ with

$$|u(x_1) - u(x_2)| \leq \mu_h |x_1 - x_2|, \quad \forall x_1, x_2 \in [-1, 1].$$

Thus, $y = u(x)$ is the intersection of the horizontal strips U^k, $k = 1, 2, \ldots$. ☐

Lemma 7.6   Assume that $0 \leq \mu_h \mu_v < 1$. Then each μ_h-horizontal curve and each μ_v-vertical curve intersect at a unique point.

Proof. It suffices to prove that the equation

$$\begin{cases} y = u(x) \\ x = v(y) \end{cases}$$

has a unique solution.

   Substituting the first equation into the second, we have

$$x = v(u(x)).$$

Since

$$|v(u(x_1)) - v(u(x_2))| \leq \mu_v |u(x_1) - u(x_2)|$$
$$\leq \mu_v \mu_h |x_1 - x_2|, \qquad \forall x_1, x_2 \in [-1, 1],$$

and $0 \leq \mu_h \mu_v < 1$, the function $v \circ u \colon [-1, 1] \to [-1, 1]$ has a unique fixed point which is the unique intersection point of the two curves. $\qquad\square$

In the following, assume that $\mu_h = \mu_v = \mu < 1$. By Lemma 7.6, a μ-horizontal curve $y = u(x)$ and a μ-vertical curve $x = v(y)$ has a unique intersecting point, denoted by $z = (x, y)$. Let

$$\|u\| = \max_{-1 \leq x \leq 1} \{|u(x)|\},$$
$$\|v\| = \max_{-1 \leq x \leq 1} \{|v(x)|\},$$
$$|z| = \max\{|x|, |y|\}.$$

Lemma 7.7 Let $z_j = (x_j, y_j)$ be the intersecting point of a μ-horizontal curve u_j and a μ-vertical curve v_j, $j = 1, 2$. Then we have

$$|z_1 - z_2| \leq \frac{1}{1 - \mu} \max\{\|u_1 - u_2\|, \|v_1 - v_2\|\}.$$

Proof. We have

$$\begin{aligned}
|x_1 - x_2| &= |v_1(y_1) - v_2(y_2)| \\
&= |v_1(y_1) - v_1(y_2)| + |v_1(y_2) - v_2(y_2)| \\
&\leq \mu|y_1 - y_2| + \|v_1 - v_2\| \\
&\leq \mu|z_1 - z_2| + \|v_1 - v_2\|, \\
|y_1 - y_2| &= |u_1(x_1) - u_2(x_2)| \\
&= |u_1(x_1) - u_1(x_2)| + |u_1(x_2) - x_2(x_2)| \\
&\leq \mu|x_1 - x_2| + \|u_1 - u_2\| \\
&\leq \mu|z_1 - z_2| + \|u_1 - u_2\|.
\end{aligned}$$

Thus,

$$|z_1 - z_2| = \mu|z_1 - z_2| + \max\{\|u_1 - u_2\|, \|v_1 - v_2\|\}$$
$$|z_1 - z_2| \leq \frac{1}{1 - \mu} \max\{\|u_1 - u_2\|, \|v_1 - v_2\|\}. \qquad (7.4)$$

$\qquad\square$

Corollary 7.8   Let V be a μ-vertical strip composed of two μ-vertical curves v_1 and v_2. A μ-horizontal curve u intersects with v_1 and v_2 at z_1 and z_2, respectively. Then

$$|z_1 - z_2| \le \frac{1}{1 - \mu}\theta(V).$$

Similarly, if U is a μ-horizontal strip composed of two μ-horizontal curves u_1 and u_2. A μ-vertical curve v intersects with u_1 and u_2 at z_1 and z_2, respectively. Then

$$|z_1 - z_2| \le \frac{1}{1 - \mu}\theta(U).$$

Proof. It follows from Lemma 7.7 by taking $u_1 = u_2$ and $v_1 = v_2$, respectively.   □

Let $\psi: Q \to \mathbb{R}^2$ be a homeomorphism. Let $U_0, \ldots, U_{N-1}$ be a collection of disjoint μ-horizontal strips; $V_0, \ldots, V_{N-1}$ be a collection of disjoint μ-vertical strips.

We make the following assumptions on ψ:

(A1) For $j = 0, \ldots, N - 1$,

$$\psi(U_j) = V_j,$$

and the horizontal boundaries of U_j are mapped onto the horizontal boundaries of V_j, the vertical boundaries of U_j are mapped onto the vertical boundaries of V_j.

(A2) Let U be a μ-horizontal strip contained in $\bigcup_{j=0}^{N-1} U_j$. Then

$$\tilde{U}_k = \psi^{-1}(V_k \cap U)$$

is a μ-horizontal strip, which is contained in U_k. Furthermore, there exists $\nu: \ 0 < \nu < 1$ such that

$$\theta(\tilde{U}_k) \le \nu\theta(U_k).$$

The same holds true about a μ-vertical strip V in the vertical direction.

Under assumptions (A1) and (A2), we can prove that ψ has a shift invariant set of order N with respect to the two-sided shift.

Similar to the approach in Section 7.1, for

$$s_{-k}, \ldots, s_{-1}; s_0, s_1, \ldots, s_k \in \{0, 1, \ldots, N - 1\} = S(N),$$

we define $U_{s_0 \ldots s_k}$ and $V_{s_{-k} \ldots s_{-2} s_{-1}}$ inductively as follows

$$\begin{aligned}
U_{s_0 \ldots s_k} &= \psi^{-1}(V_{s_0} \cap U_{s_1 \ldots s_k}) = U_{s_0} \cap \psi^{-1}(U_{s_1 \ldots s_k}) \\
&= U_{s_0} \cap \psi^{-1}(U_{s_1}) \cap \cdots \cap \psi^{-k}(U_{s_k}); \\
V_{s_{-k} \ldots s_{-2} s_{-1}} &= \psi(U_{s_{-1}} \cap V_{s_{-k} \ldots s_{-2}}) = V_{s_{-1}} \cap \psi(V_{s_{-k} \ldots s_{-2}}) \\
&= V_{s_{-1}} \cap \psi(V_{s_{-2}}) \cap \cdots \cap \psi^{k-1}(V_{s_{-k}}).
\end{aligned}$$

From assumptions (A1) and (A2), it is easy to show that $U_{s_0...s_k}$ is a μ-horizontal strip contained in $U_{s_0...s_{k-1}}$, and that $V_{s_{-k}...s_{-1}}$ is a μ-vertical strip contained in $V_{s_{-k+1}...s_{-1}}$.

Similar to Lemma 7.1, we have the following.

Lemma 7.9   Under assumptions (A1) and (A2), we have

(i) $\psi(U_{s_0...s_k}) = V_{s_0} \cap U_{s_1...s_k}$, $\psi(V_{s_{-k}...s_{-2}s_{-1}}) \cap V_{s_0} = V_{s_{-k}...s_{-1}s_0}$;

(ii) $\theta(U_{s_0...s_k}) < \nu\theta(U_{s_1...s_k}) < \nu^k$, $\theta(V_{s_{-k}...s_{-2}s_{-1}}) < \nu\theta(V_{s_{-k}...s_{-2}}) < \nu^{k-1}$. $\qquad\square$

Next, For $s = (\ldots, s_{-2}, s_{-1}; s_0, s_1, s_2, \ldots) \in \sum_N$, we define

$$U(s) = \bigcap_{j=0}^{\infty} \psi^{-j}(U_{s_j})$$

$$= \bigcap_{k=0}^{\infty} U_{s_0 s_1...s_k} \left(= \bigcap_{k=1}^{\infty} U_{s_0 s_1...s_k} \right),$$

$$V(s) = \bigcap_{j=1}^{\infty} \psi^{j-1}(V_{s_{-j}})$$

$$= \bigcap_{k=1}^{\infty} V_{s_{-k}...s_{-1}} \left(= \bigcap_{k=2}^{\infty} V_{s_{-k}...s_{-1}} \right).$$

Lemma 7.10   We have

(i) $\psi(V(s) \cap U(s)) = V(\sigma(s)) \cap U(\sigma(s))$;

(ii) $\sharp(V(s) \cap U(s)) = 1$.

Proof. Part (i) is similar to the proof for part (i) in Lemma 7.2. We leave it to the readers as an exercise.

Part (ii) follows from the fact that the μ-horizontal curve $U(s)$ and the μ-vertical curve $V(s)$ have a unique intersecting point by Lemma 7.6. $\qquad\square$

Finally, we define

$$\triangle = \bigcap_{s \in \Sigma_N} (V(s) \cap U(s)). \tag{7.5}$$

Theorem 7.11   The set $\triangle$ in (7.5) is a compact invariant set of ψ, and $\psi|_\triangle$ is topologically conjugate to the two-sided shift $\sigma : \sum_N \to \sum_N$. Therefore, ψ has a shift invariant set of order N with respect to the two-sided shift.

Proof. By identifying a singleton set with the singleton itself therein, we define the map $h\colon \sum_N \to \triangle$ as follows:

$$h(s) = V(s) \cap U(s).$$

It remains to show that h is a topological conjugacy of ψ and σ. Here we just check the continuity of h. The proof of the remainders is similar to that of Theorem 6.52.

For the continuity, let $s, t \in \sum_N$ with

$$d(s, t) < \frac{1}{2^k}.$$

Then

$$h(s), h(t) \in V_{s_{-k} \ldots s_{-1}} \cap U_{s_0 \ldots s_k}.$$

It follows from Lemma 7.7 that

$$|h(s) - h(t)| \leq \frac{1}{\mu} \max\{\theta(V_{s_{-k} \ldots s_{-1}}), \theta(U_{s_0 \ldots s_k})\}$$

$$\leq \frac{\nu^{k-1}}{1 - \mu}. \tag{7.6}$$

$\square$

In the following exercises, we offer some alternative and easily checkable conditions which substitute the condition (A2).

Exercise 7.12   Consider the Henon map $F\colon \mathbb{R}^2 \to \mathbb{R}^2$:

$$F(x, y) = (1 + y - ax^2, bx).$$

Fix $a = 1.4$ and $b = 0.3$. Draw a rectangle in the (x, y) plane such that its image under the Henon map intersect the rectangle two times. $\square$

Exercise 7.13   Consider the Henon map as in Exercise 7.12. Show that it has a Smale horseshoe.$\square$

Exercise 7.14   Construct a smooth function $\beta\colon \mathbb{R} \to \mathbb{R}$ with the following properties:

(i) $\beta(t) > 0$, if $|t| < 1$, and $\beta(t) = 0$, if $|t| \geq 1$;

(ii) $\beta(-t) = \beta(t)$, $\forall t \in \mathbb{R}$;

(iii) $\int_{-\infty}^{+\infty} \beta(t)dt = 1$. $\square$

Exercise 7.15   Let $v\colon [-1, 1] \to \mathbb{R}$ satisfy

$$|v(t_1) - v(t_2)| \leq \mu |t_1 - t_2|, \qquad \forall t_1, t_2 \in [-1, 1],$$
$$|v(t)| \leq 1, \qquad \forall t \in [-1, 1],$$

Prove that for any $\varepsilon > 0$, there exists a *smooth* function $\tilde{v}\colon \mathbb{R} \to \mathbb{R}$ such that

$$|\tilde{v}(t_1) - \tilde{v}(t_2)| \leq \mu |t_1 - t_2|, \qquad \forall t_1, t_2 \in \mathbb{R},$$
$$|\tilde{v}(t)| \leq 1, \qquad \forall t \in \mathbb{R},$$
$$|\tilde{v}(t) - v(t)| < \varepsilon, \qquad \forall t \in [-1, 1]. \qquad \square$$

In addition to condition (A1), we introduce the following condition: Assume that $\psi\colon P \to \psi(P)(\subset \mathbb{R}^2)$ is a diffeomorphism that satisfies

(A3) For any $p \in \bigcup_{j=0}^{N-1} U_j$, the map

$$\begin{pmatrix} \xi_1 \\ \eta_1 \end{pmatrix} = D\psi_p \begin{pmatrix} \xi_0 \\ \eta_0 \end{pmatrix}, \qquad \forall \begin{pmatrix} \xi_0 \\ \eta_0 \end{pmatrix} \in \mathbb{R}^2,$$

satisfies the property that if $|\xi_0| \leq \mu |\eta_0|$, then

$$|\xi_1| \leq \mu |\eta_1|, \qquad |\eta_1| \geq \mu^{-1} |\eta_0|.$$

Likewise, for any $q \in \bigcup_{j=0}^{N-1} V_j$, the map

$$\begin{pmatrix} \xi_0 \\ \eta_0 \end{pmatrix} = D\psi_q^{-1} \begin{pmatrix} \xi_1 \\ \eta_1 \end{pmatrix}, \qquad \forall \begin{pmatrix} \xi_1 \\ \eta_1 \end{pmatrix} \in \mathbb{R}^2,$$

satisfies the property that if $|\eta_1| \leq \mu |\xi_1|$, then

$$|\eta_0| \leq \mu |\xi_0|, \qquad |\xi_0| \geq \mu^{-1} |\xi_1|. \qquad \square$$

Exercise 7.16   Let $\psi\colon P \to \mathbb{R}^2$ be a diffeomorphism from P to $\psi(P)$ that satisfies assumptions (A1) and (A3). Let $\gamma \subset \bigcup_{j=0}^{N-1} V_j$ be a μ-vertical curve and $\delta \subset \bigcup_{j=0}^{N-1} U_j$ be a μ-horizontal curve. Prove that the image $\psi(\hat{\gamma})$ of $\hat{\gamma} = U_k \cap \gamma$ is a μ-vertical curve, and the image $\psi(\check{\delta})$ of $\check{\delta} = V_l \cap \delta$ is a μ-horizontal curve. $\qquad \square$

Exercise 7.17   Prove that under the assumptions of Exercise 7.16 with $0 < \mu < \frac{1}{2}$, assumption (A2) holds. $\qquad \square$

NOTES FOR CHAPTER 7

The Smale horseshoe, due to Stephen Smale [64] in 1963, is the earliest, most prominent example of higher-dimensional chaotic map. It is crucial in the understanding of how and why certain dynamics becomes chaotic, and then in developing the analysis to rigorously prove the occurrence of chaos. The horseshoe map is a geometrical and global concept. Its invariant set is a Cantor set, with infinitely many periodic points and uncountably many non-periodic orbits and yet it is "structurally stable."

Our treatment in this chapter is based mostly on that in Wiggins [69, Chapter 4], Zhang [74] and Zhou [75]. Section 7.1 studies the standard Smale horseshoe by symbolic dynamics by showing that the horseshoe diffeomorphism is topologically conjugate to a full shift on two-symbols on an invariant Cantor set. Section 7.2 does the general horseshoe with the Conley-Moser condition (Moser [56]), leading to a topological conjugacy to a full shift on N-symbols. Further generalizations to dimensions higher than two can be found in Wiggins [68].

Theorem 7.2 in Section 7.2 is adopted from Zhang [74].

CHAPTER 8

Fractals

We begin this chapter by giving some simple constructions.

8.1 EXAMPLES OF FRACTALS

Example 8.1 (The classical Cantor set) The Cantor ternary set constructed in Example 5.12 is a standard example of a fractal, the subject of this chapter. □

Example 8.2 (The Sierpinski gasket) Let S_0 be a triangle with sides of unit-length. Connecting the middle point of each side, we obtain four triangles, each side has equal-length $\frac{1}{2}$. Deleting the interior of the middle one, the remainder part is denoted by S_1, which is composed of three triangles. Repeating the procedure for each triangle in S_1, we have S_2. Continuing this procedure, we obtain $S_0, S_1, S_2, \ldots$. The nonempty set

$$\mathcal{G} = \bigcap_{i=0}^{\infty} S_i$$

is the *Sierpinski gasket*. See Fig. 8.1. It is also called the Sierpinski triangle or the Sierpinski Sieve. The fractal is named after the Polish mathematician Wacław Sierpiński, who described it in 1915. Cf. also [72]. □

Example 8.3 (The Koch Curve) The Koch curve, due to the Swedish mathematician Helge von Koch, is constructed by first drawing an equilateral triangle in $\mathbb{R}^2$, then recursively alter each line segment as follows:

(1) divide each side of the triangle into three segments of equal length;

(2) draw an outward pointing equilateral triangle that has the middle segment from step (1) as its base;

(3) remove the side that is the base of the triangle from step (2).

Performing just one iteration of this process, one obtains the result in the shape of the Star of David.

By iterating indefinitely, the limit is the *Koch curve*, which is also called the Koch snowflake, the Koch star or the Koch island. See Fig. 8.2, and [71]. □

Figure 8.1: The Sierpinski gasket (Example 8.2).

Figure 8.2: The Koch curve.

8.2 HAUSDORFF DIMENSION AND THE HAUSDORFF MEASURE

Let A be a set in $\mathbb{R}^N$. The *diameter of A* is defined as

$$|A| = \sup\{|x - y| \mid x, y \in A\}.$$

Let $E \subset \mathbb{R}^N$. We call $\alpha = \{U_i \mid i > 0\}$ a (countable) cover of E if $E \subset \bigcup_{i>0} U_i$. Let $\delta > 0$. The collection α is called a δ-cover of E if

$$E \subset \bigcup_{i>0} U_i, \qquad 0 < |U_i| \le \delta, \qquad \forall i > 0.$$

Let $s \ge 0, \delta > 0$, define

$$\mathcal{H}_\delta^s(E) = \inf \sum_{i=1}^{\infty} |U_i|^s. \tag{8.1}$$

Here the infimum is taken over all the δ-covers of E. It can be checked that $\mathcal{H}_\delta^s$ is a metric outer measure on $\mathbb{R}^N$.

Define

$$\mathcal{H}^s(E) = \lim_{\delta \to 0} \mathcal{H}^s_\delta(E) = \sup_{\delta \to 0} \mathcal{H}^s_\delta(E).$$

The limit exists, but may be infinite, since $\mathcal{H}^s_\delta$ is decreasing as a function of δ. $\mathcal{H}^s$ is also a metric outer measure. The restriction of $\mathcal{H}^s$ to the σ-field of $\mathcal{H}^s$-measurable sets is called *the Hausdorff s-dimensional measure* of the set E.

Note that an equivalent definition of the Hausdorff measure is obtained if the infimum in (8.1) is taken over all δ-covers of E by convex sets rather than by arbitrary sets. If the infimum is taken over δ-covers of open (closed) sets, a different value of $\mathcal{H}^s_\delta$ may result, but the value of the limit $\mathcal{H}^s$ is the same.

Exercise 8.4   Prove that for any E, $\mathcal{H}^s(E)$ is non-increasing as s increases from 0 to ∞. Furthermore, if $s < t$, then

$$\mathcal{H}^s_\delta(E) \geq \delta^{s-t} \mathcal{H}^t_\delta(E),$$

(which implies that if $\mathcal{H}^t(E)$ is positive, then $\mathcal{H}^s(E)$ is infinite). $\qquad\qquad\qquad\square$

Thus, from Exercise 8.4, there is a unique value, denoted by $\dim_{\mathcal{H}}(E)$, called the Hausdorff dimension of E, such that

$$\begin{aligned}
\mathcal{H}^s(E) &= \infty, \quad \text{if } 0 \leq s < \dim_{\mathcal{H}}(E), \\
\mathcal{H}^s(E) &= 0, \quad \text{if } \dim_{\mathcal{H}}(E) < s < \infty.
\end{aligned}$$

Alternatively, we also have

$$\dim_{\mathcal{H}}(E) = \inf\{s: \ \mathcal{H}^s(E) = 0\} = \sup\{s: \ \mathcal{H}^s(E) = \infty\}.$$

In the following, denote $s = \dim_{\mathcal{H}}(E)$. In general, for a set E, the Hausdorff dimension $s = \dim_{\mathcal{H}}(E)$ of E may not be an integer or even a fraction.

A measurable set $E \subset \mathbb{R}^N$ is said to be an s-set if $0 < \mathcal{H}^s(E) < \infty$. It is obvious that $\dim_{\mathcal{H}}(E) = s$, provided that E is an s-set. In the chapter, we will see many examples of s-sets.

Some elementary properties of the Hausdorff dimension and the Hausdorff measure are listed in the following.

Lemma 8.5

(i) Monotonicity: Let E, $F \subset \mathbb{R}^N$. If $E \subset F$, then $\mathcal{H}^s(E) \leq \mathcal{H}^s(E)$ and $\dim_{\mathcal{H}}(E) \leq \dim_{\mathcal{H}}(F)$.

(ii) Countable stability: Let $\{F_i\}_{i=1}^\infty$ be a countable sequence of sets in $\mathbb{R}^N$. Then

$$\dim_{\mathcal{H}}\left(\bigcup_{i>0} F_i\right) = \sup_{i>0}\{\dim_{\mathcal{H}}(F_i)\}.$$

(iii) If F is a countable set in $\mathbb{R}^N$, then $\dim_{\mathcal{H}}(F) = 0$. If F is a finite set, then $\mathcal{H}^0(F) = \text{card}\{F\}$.

(iv) Let $E \subset \mathbb{R}^N$. If $f\colon F \to \mathbb{R}^N$ is Hölder continuous, i.e., there exist constants $c > 0$ and $\alpha\colon 0 < \alpha \le 1$ such that

$$|f(x) - f(y)| \le c|x - y|^{\alpha}, \qquad \forall x, y \in F,$$

then

$$\mathcal{H}^{s/\alpha}(f(F)) \le c^{s/\alpha}\mathcal{H}^s(F), \qquad \dim_{\mathcal{H}}(f(F)) \le \frac{1}{\alpha}\dim_{\mathcal{H}}(F).$$

(v) In particular, if f is Lipschitz with some Lipschitz constant $c > 0$, then

$$\mathcal{H}^s(f(F)) \le c^s\mathcal{H}^s((F)), \qquad \dim_{\mathcal{H}}(f(F)) \le \dim_{\mathcal{H}}(F).$$

Furthermore, if f is bi-Lipschitz, i.e., there exist $0 < c_1 < c$ such that

$$c_1|x - y| \le |f(x) - f(y)| \le c|x - y|, \qquad \forall x, y \in F,$$

then

$$c_1^s\mathcal{H}^s((F)) \le \mathcal{H}^s(f(F)) \le c^s\mathcal{H}^s((F)), \qquad \dim_{\mathcal{H}}(f(F)) = \dim_{\mathcal{H}}(F).$$

(vi) Let $E \subset \mathbb{R}^N$ and $\lambda > 0$. Denote $\lambda E = \{\lambda x \mid x \in E\}$. Then

$$\mathcal{H}^s(\lambda E) = \lambda^s\mathcal{H}^s(E).$$

Proof.   (i) This follows directly from the definition.

(ii) By (i), we have

$$\dim_{\mathcal{H}}\left(\bigcup_{i>0} F_i\right) \ge \dim_{\mathcal{H}}(F_i), \qquad \forall i > 0.$$

On the other hand, if $s > \dim_{\mathcal{H}}(F_i)$, $\forall i > 0$. then $\mathcal{H}^s(F_i) = 0$, $\forall i > 0$ by definition. it follows that $\mathcal{H}^s\left(\bigcup_{i>0} F_i\right) = 0$ by the additivity of measure, Thus, we have

$$\dim_{\mathcal{H}}\left(\bigcup_{i>0} F_i\right) \le \sup_{i>0}\{\dim_{\mathcal{H}}(F_i)\}.$$

(iii) Obviously, $\mathcal{H}^0(F) = 1$ when F contains only a single point. The first part follows from the countable stability in part (ii). The second part can be proved directly from the definition.

(iv) Let $\{U_i\}$ be a δ-cover of F. By assumption,

$$|f(F \cap U_i)| \leq c|U_i|^\alpha,$$

which implies that $\{f(F \cap U_i)\}$ is a $c\delta^\alpha$-cover of $f(E)$. Therefore,

$$\sum_{i>0} |f(F \cap U_i)|^{s/\alpha} \leq c^{s/\alpha} \sum_{i>0} |U_i|^s,$$

$$\mathcal{H}^{s/\alpha}_{c\delta^\alpha}(f(F)) \leq c^{s/\alpha} \mathcal{H}^s_\delta(F).$$

Letting $\delta \to 0$, (so $c\delta^\alpha \to 0$), we get

$$\mathcal{H}^{s/\alpha}(f(F)) \leq c^{s/\alpha} \mathcal{H}^s(F), \qquad \dim_{\mathcal{H}}(f(F)) \leq \frac{1}{\alpha} \dim_{\mathcal{H}}(F).$$

(v) is a particular case of (iv) that $\alpha = 1$.

(vi) Let $\{U_i\}$ be a δ-cover of E. Then $\{\lambda U_i\}$ is a $\lambda\delta$-cover of λE. So

$$\mathcal{H}^s_{\lambda\delta}(\lambda E) \leq \sum_i |\lambda U_i|^s = \lambda^s \sum_i |U_i|^s \leq \lambda^s \mathcal{H}^s_\delta(E).$$

Letting $\delta \to 0$, we have

$$\mathcal{H}^s(\lambda E) \leq \lambda^s \mathcal{H}^s(E).$$

The converse inequality follows by taking $\frac{1}{\lambda}$ instead of λ.

$\square$

There are many other properties about the Hausdorff dimension and the Hausdorff measure. For interested readers, see [27, 28, 42, 67].

Finally, in this section, we state the following relationship between the N-dimensional Lebesgue measure $\mathcal{L}^N$ and N-dimensional Hausdorff measure. For the proof, see [29, Theorem 1.12, p. 13].

Lemma 8.6 If $E \subset \mathbb{R}^N$, then

$$\mathcal{L}^N(E) = c_N \mathcal{H}^N(E),$$

where $c_N = \pi^{\frac{1}{2}N}/2^N(\frac{1}{2}N)!$ is the volume of the ball in $\mathbb{R}^N$ with diameter 1.

$\square$

8.3 ITERATED FUNCTION SYSTEMS (IFS)

This section presents an easy way to generate complicated sets, the so-called *self-similar sets*, which are generated by iterated function systems.

A map $S\colon \mathbb{R}^N \to \mathbb{R}^N$ is called a *contraction* if $|S(x) - S(y)| \leq c|x - y|$ for all x, $y \in \mathbb{R}^N$, for some $c\colon 0 < c < 1$. We call the infimum of such c the *ratio of contraction*. Furthermore, if $|S(x) - S(y)| = c|x - y|$ for all x, $y \in \mathbb{R}^N$ for some $c < 1$, then S is called a *similitude*. Geometrically, a similitude maps every subset of $\mathbb{R}^N$ to a similar set. Thus, it is a composition of a dilation with a rotation, a translation and perhaps a reflection as well. Such a map can be written as

$$S(x) = cQx + r, \qquad x, r \in \mathbb{R}^N,$$

where Q is an *orthogonal matrix*.

An iterated function system on $\mathbb{R}^N$ is a finite collection of contractions $\{S_0, S_1, \ldots, S_{m-1}\}$ with $m > 1$. We refer to such a system as an *IFS*. A non-empty compact set E in $\mathbb{R}^N$ is said to be an invariant set of the IFS if

$$E = \bigcup_{i=0}^{m-1} S_i(E).$$

In the following, we show that an IFS has a unique invariant set. Let $\mathcal{C}(\mathbb{R}^N)$ denote the class of all non-empty compact sets of $\mathbb{R}^N$. For any E, $F \in \mathcal{C}(\mathbb{R}^N)$, define the distance between E and F:

$$\rho_H(E, F) = \max\{\max_{x \in E} d(x, F), \max_{y \in F} d(y, E)\}, \tag{8.2}$$

where, for a non-empty closed set S of $\mathbb{R}^N$, $d(x, S)$ is the distance from a point x to the set S:

$$d(x, S) = \min_{y \in S}\{|x - y|\}.$$

Because the set is closed, the minimum is attained. We can prove that ρ_H is a metric on $\mathcal{C}(\mathbb{R}^N)$ (Exercise), which is called *the Hausdorff distance* and that $(\mathcal{C}(\mathbb{R}^N), \rho_H)$ become a complete metric space.

Since

$$\max_{x \in E}\{d(x, F)\} = \min\{\delta \geq 0\colon\ E \subset F(\delta)\},$$
$$\max_{y \in F}\{d(y, E)\} = \min\{\delta \geq 0\colon\ F \subset E(\delta)\},$$

an equivalent definition of ρ_H is

$$\rho_H(E, F) = \inf\{\delta > 0\colon\ E \subset F(\delta),\ \text{and}\ F \subset E(\delta)\}. \tag{8.3}$$

Here

$$E(\delta) = \{x \in \mathbb{R}^N\colon\ d(x, E) \leq \delta\}, \quad F(\delta) = \{x \in \mathbb{R}^N\colon\ d(x, F) \leq \delta\}$$

are δ-closed neighborhoods of E and F, respectively.

Lemma 8.7 Assume that S is a single contraction on $\mathbb{R}^N$ with contraction ratio $c < 1$. Then S induces a contraction on $\mathcal{C}(\mathbb{R}^N)$ with the same contraction ratio c.

Proof. Let E and F be two sets in $\mathcal{C}(\mathbb{R}^N)$. For any $\rho > \rho_H(E, F)$, by (8.3), it follows that $E \subset F(\rho)$ and $F \subset E(\rho)$. That is,

$$d(x, F) < \rho, \qquad \forall x \in E,$$

which implies

$$d(S(x), S(F)) \le cd(x, F) < c\rho, \qquad \forall x \in E.$$

Thus, $S(E) \subset S(F)(c\rho)$. By the same way, we have $S(F) \subset S(E)(c\rho)$. Again by (8.3), we have

$$\rho_H(S(E), S(F)) \le c\rho.$$

By the arbitrariness of $\rho > \rho_H(E, F)$, we get

$$\rho_H(S(E), S(F)) \le c\rho_H(E, F).$$

Thus, a single contraction induces a contraction on $\mathcal{C}(\mathbb{R}^N)$. □

Theorem 8.8 Given an IFS $\{S_0, S_1, \ldots, S_{m-1}\}$ with contraction ratios $0 < c_i < 1$ of S_i for $i = 0, 1, \ldots, m - 1$, there exists a unique invariant set.

Proof. The induced map $\mathcal{S}$ on the complete metric space $(\mathcal{C}(\mathbb{R}^N), \rho_H)$ by the IFS is defined as

$$\mathcal{S}(E) = \bigcup_{i=0}^{m-1} S_i(E), \qquad \forall E \in \mathcal{C}(\mathbb{R}^N).$$

Any fixed point of $\mathcal{S}$ is an invariant set of the IFS. To prove the theorem, it suffices to prove that the map $\mathcal{S}$ on $\mathcal{C}(\mathbb{R}^N)$ has a unique fixed point. This is done if we can show that $\mathcal{S}$ is a contraction, since the space $\mathcal{C}(\mathbb{R}^N)$ is complete.

For any $E, F \in \mathcal{C}(\mathbb{R}^N)$, we claim

$$\begin{aligned}
\rho_H(\mathcal{S}(E), \mathcal{S}(F)) &= \rho_H\left(\bigcup_{i=0}^{m-1} S_i(E), \bigcup_{i=0}^{m-1} S_i(F)\right) \\
&\le \max_{0 \le i \le m-1} \rho_H(S_i(E), S_i(F)).
\end{aligned} \tag{8.4}$$

In fact, if $\delta > 0$ such that

$$S_i(E)(\delta) \supset S_i(F), \qquad \forall i = 0, 1, \ldots, m - 1,$$

then

$$\left(\bigcup_{i=0}^{m-1} S_i(E)\right)(\delta) \supset \bigcup_{i=0}^{m-1} S_i(\delta) \supset \bigcup_{i=0}^{m-1} S_i(F).$$

The same is also true by exchanging the roles of E and F above. This proves our claims.

It follows from (8.4) and Lemma 8.7 that

$$\rho_H(\mathcal{S}(E), \mathcal{S}(F)) \leq \max_{0 \leq i \leq m-1} c_i \rho_H(E, F).$$

That is, $\mathcal{S}$ is a contraction on $C(\mathbb{R}^N)$ with contraction ratio $c = \max\limits_{0 \leq i \leq m-1} c_i$ < 1. $\qquad\square$

Definition 8.9 (Self-similarity) The unique invariant set E warranted by Theorem 8.8 is called the self-similar set generated by the family of similitudes $\{S_0, S_1, \ldots, S_{m-1}\}$.

To obtain the invariant set, we can take the IFS as acting on the sets. In fact, by the proof of Theorem 8.8, S is a contraction on $C(\mathbb{R}^N)$. So starting with any non-empty compact set F_0, the sequence of sets $F_n = S(F_{n-1})$ converges to the fixed point in $C(\mathbb{R}^N)$. It is the unique compact set in $C(\mathbb{R}^N)$ that is the attractor and invariant of the IFS.

We now consider the construction of the invariant set of an IFS. We need the following.

Lemma 8.10 Let $\{S_0, S_1, \ldots, S_{m-1}\}$ be an IFS with each S_i's contraction ratio $0 < c_i < 1$. Denote

$$R = \max_{0 \leq i \leq m-1} \left\{ \frac{\|S_i(0)\|}{1 - c_i} \right\},$$
$$\bar{B}(R) = \{x \in \mathbb{R}^N \mid |x| \leq R\}.$$

Then this ball $\bar{B}(R)$ is positively invariant under the IFS, or

$$S_i(\overline{B}(R)) \subset \overline{B}(R).$$

Proof. By the choice of R, $|S_i(0)| \leq R(1 - c_i)$. For $x \in \overline{B}(R)$,

$$\begin{aligned}
|S_i(x)| &\leq |S_i(x) - S_i(0)| + |S_i(0)| \\
&\leq c_i|x - 0| + |S_i(0)| \\
&\leq c_i R + R(1 - c_i) \\
&= R.
\end{aligned}$$

This shows that the ball $\overline{B}(R)$ is positively invariant under all of the S_i. $\qquad\square$

Let $\sum_m^+$ be the symbolic space as defined in Section 6.3. Let F be a positively compact invariant set under all the S_i, $i = 0, \ldots, m - 1$. For any $s = (s_0 s_1 \ldots) \in \sum_m^+$, we have

$$S_{s_0} \ldots S_{s_{k+1}}(F) \subset S_{s_0} \ldots S_{s_k}(F), \qquad \forall k \geq 0,$$

and

$$|S_{s_0} \ldots S_{s_k}(F)| \leq c_{s_0} \ldots c_{s_k} |F| \to 0, \quad \text{as} \quad k \to \infty.$$

Thus, the set $\bigcap_{k=0}^{\infty} S_{s_0} \ldots S_{s_k}(F)$ contains only a singleton. Denote

$$\{x_s\} = \bigcap_{k=0}^{\infty} S_{s_0} \ldots S_{s_k}(F),$$

and

$$E = \bigcup_{s \in \sum_m^+} \{x_s\}. \tag{8.5}$$

Theorem 8.11 Let $\{S_0, S_1, \ldots, S_{m-1}\}$ be an IFS with contraction ratio $0 < c_i < 1$ for S_i, $i = 0, 1, \ldots, m - 1$, and E be defined as in (8.5). Then

(i) E is the unique invariant set of the IFS.

(ii) Let $(\sum_m^+, \sigma^+)$ be the one-sided shift defined in Section 6.3. Define $\alpha \colon E \to E$ by

$$\alpha(x_s) = x_{\sigma^+(s)}, \qquad x_s \in E.$$

Then σ^+ is semiconjugate to α with a semiconjugacy $h \colon \sum_m^+ \to E$:

$$h(s) = x_s, \qquad \forall s = (s_0 s_1 \ldots) \in \sum_m^+,$$

i.e.,

$$h\sigma^+ = \alpha h. \tag{8.6}$$

Proof. (i) By the definition of E, it is obvious that

$$S_i(E) = E, \qquad \forall i = 0, \ldots m - 1,$$

which implies

$$E = \bigcup_{i=0}^{m-1} S_i(E).$$

This shows that E is the unique invariant set under the IFS.

(ii) From the definition of α, it is easy to show that

$$h\sigma^+ = \alpha h.$$

It remains to show that $h\colon \sum_m^+ \to E$ is continuous and onto. We leave it as an exercise.

$\square$

Let $\sum_m^*$ denote the set of all the transitive points in $\sum_m^+$ as in Proposition 6.20. By Theorem 8.11, we have the following.

Corollary 8.12   Let $\{S_0, S_1, \ldots S_{m-1}\}$ be an IFS with contraction ratio $0 < c_i < 1$ for S_i, $i = 0, 1, \ldots, m-1$, and E be the unique invariant set. Then we have the following.

(i) For any $s \in \sum_m^*$ with $s = (s_0 s_1 s_2 \ldots)$ and $x_0 \in E$, the orbit

$$\{S_{s_j} \ldots S_{s_0}(x_0)\}_{j=0}^{\infty}$$

is dense in E.

(ii) For any $s \in \sum_m^*$ and $x_0 \in \mathbb{R}^N$, the closure of the orbit

$$\{S_{s_j} \ldots S_{s_0}(x_0)\}_{j=0}^{\infty}$$

contains E.

$\square$

Intuitively, a set is *self-similar* (according to Definition 8.9) if an arbitrary small piece of it can be magnified to give the whole set. Many of the classical fractal sets are self-similar. The three classical examples are the Cantor set, the Koch curve and the Sierpinski gasket. Self-similar sets can be generated by IFS of similitudes with the *open set condition*. An IFS $\{S_0, \ldots, S_{m-1}\}$ is called an IFS of similitudes if each contraction S_i is a similitude. Furthermore, we say that the open set condition the *open set condition* holds for $\{S_0, \ldots, S_{m-1}\}$ if there exists a bounded set V in $\mathbb{R}^N$ such that

$$\mathcal{S}(V) = \bigcup_{i=0}^{m-1} S_i(V) \subset V$$

with this union a disjoint union. It is easy to see that the unique attractor of an IFS of similitudes satisfying the open set condition is a self-similar set, which is also called a *self-similar set* with the open set condition.

Example 8.13   The classical Cantor set $\mathcal{C}$ is generated by S_0, S_1:

$$S_0(x) = \frac{1}{3}x, \qquad S_1(x) = \frac{1}{3}x + \frac{2}{3}.$$

Let $V = (0, 1)$. Then

$$\mathcal{S}(V) = S_0(V) \cup S_1(V) \subset V, \qquad S_0(V) \cap S_1(V) = \emptyset.$$

The open set condition holds. It is obvious that

$$C = \bigcap_{n=0}^{\infty} \mathcal{S}^n(\overline{V}).$$

Thus, the classical Cantor set is a self-similar set with the open set condition. $\square$

Example 8.14 The Sierpinski gasket $\mathcal{G}$. A family of similitudes on $\mathbb{R}^2$ are defined as

$$S_0(x) = \frac{1}{2}x, \qquad x = (x_1, x_2),$$
$$S_1(x) = \frac{1}{2}x + \left(\frac{1}{2}, 0\right)$$
$$S_2(x) = \frac{1}{2}x + \left(\frac{1}{4}, \frac{\sqrt{3}}{4}\right).$$

Then $\{S_0, S_1, S_2\}$ satisfies the open set condition with V a filled-in equilateral triangle whose sides are each of length 1 and whose vertices are at $(0, 0)$, $(1, 0)$ and $(1/2, \sqrt{3}/2)$. Then

$$\mathcal{G} = \bigcap_{n=0}^{\infty} \mathcal{S}^n(\overline{V}).$$

Thus, the Sierpinski gasket $\mathcal{G}$ is self-similar with the open set condition. $\square$

The calculation of the Hausdorff dimensions of fractal sets is one of the main topics in fractal geometry. It is a difficult problem for a general fractal set. But there is an elegant result about the Hausdorff dimensions of self-similar sets with the open set condition.

Proposition 8.15 ([29, Theorem 8.6, p. 121]) Let $S_i\colon \mathbb{R}^N \to \mathbb{R}^N$, $1 \le i \le m$, be an IFS of similitudes with the open set condition. That is

(1) S_i is a similar contraction with contraction ratio c_i $0 < c_i < 1$, i.e.,

$$|S_i(x) - S_i(y)| = c_i|x - y|, \qquad \forall x, y \in \mathbb{R}^N,$$

(2) $S_1, S_2, \ldots, S_m$ satisfy the open set condition.

Then the unique invariant set E as warranted in Theorem 8.11 (which is a self-similar set satisfying the open set condition) is an s-set, where s is determined by

$$\sum_{i=1}^{m} c_i^s = 1;$$

in particular, $0 < \mathcal{H}^s(E) < \infty$. $\qquad\qquad\qquad\qquad\qquad\qquad\qquad\qquad\Box$

From Proposition 8.15, the Hausdorff dimension s of the classical Cantor set $\mathcal{C}$ satisfies the equation

$$\left(\frac{1}{3}\right)^s + \left(\frac{1}{3}\right)^s = 1,$$

which implies that

$$\dim_{\mathcal{H}}(\mathcal{C}) = s = \frac{\ln 2}{\ln 3}.$$

Through a tedious computation, it is possible to show that $\mathcal{H}^s(\mathcal{C}) = 1$ [28].

By a similar argument, we have the Sierpinski gasket $\mathcal{G}$'s dimension as

$$\dim_{\mathcal{H}}(\mathcal{G}) = \frac{\ln 3}{\ln 2}.$$

Let $s \geq 0$ and $F \subset \mathbb{R}^N$ define

$$\mathcal{P}_\delta^s(F) = \sup\left\{\sum_i |B_i|\right\},$$

where B_i is a ball centered in F and with radius less than or equal to δ (> 0) and $B_i \cap B_j = \emptyset$, ($i \neq j$). Let

$$\mathcal{P}_0^s(F) = \lim_{\delta \to 0} \mathcal{P}_\delta^s(F).$$

This limit exists since $\mathcal{P}_\delta^s(F)$ is monotone as a function of δ.

Let

$$\mathcal{P}^s(F) = \inf\left\{\sum \mathcal{P}_0^s(F_i): \ |F = \cup F_i\right\}.$$

It is easy to see that $\mathcal{P}^s(\cdot)$ is a measure on $\mathbb{R}^N$ which is called packing measure with dimension s.

The *packing dimension* of F is defined as

$$\dim_{\mathcal{P}}(F) = \sup\{s: \ \mathcal{P}^s(F) = \infty\} = \inf\{s: \ \mathcal{P}^s(F) = 0\}.$$

To finish this section, we state a result on the Hausdorff dimension of product sets.

Proposition 8.16 Let $E \subset \mathbb{R}^M$ and $F \subset \mathbb{R}^N$. Then there exists a positive constant c which depends only on s and t such that

$$\mathcal{H}^{s+t}(E \times F) \geq c\mathcal{H}^s(E)\mathcal{H}^t(F).$$

Furthermore, if one of the E and F is regular (i.e., its Hausdorff dimension is equal to its packing dimension), then

$$\dim_{\mathcal{H}}(E \times F) = \dim_{\mathcal{H}}(E) + \dim_{\mathcal{H}}(F).$$ □

For the proof, see [67, Theorem 1, p. 103 and Proposition 1, p. 106].

Exercise 8.17   Let $\rho_H(\cdot, \cdot)$ be defined as in (8.2). Prove that ρ_H is a distance on $\mathcal{C}(\mathbb{R}^N)$ and $(\mathcal{C}(\mathbb{R}^N), \rho_H)$ is a complete metric space.   □

Exercise 8.18   Prove that the map h defined in the proof of Theorem 8.11 is continuous and onto.
□

Exercise 8.19   Find the IFS of similitudes which generates the Koch curve and show that it is self-similar satisfying the open set condition.   □

Exercise 8.20   Prove that the Hausdorff dimension of the Koch curve is $\log_3 4$.

□

Exercise 8.21   For the classical Cantor set $\mathcal{C}$, show that

$$\mathcal{H}^s(\mathcal{C}) = 1,$$

where $s = \log_3 2$.   □

Exercise 8.22   Consider

$$f_\mu(x) = \begin{cases} \mu x, & 0 \le x \le 1/2, \\ \mu(1-x), & 1/2 < x \le 1, \end{cases} \qquad (\mu > 0).$$

For $\mu > 2$, let

$$X = \{x \in [0, 1] \mid f_\mu^n(x) \in [0, 1], \text{ for all } n = 0, 1, 2, \ldots\},$$

be the largest invariant set of f_μ contained in $[0, 1]$. Prove that

(i) X is a self-similar set satisfying the open set condition;

(ii) The Hausdorff dimension of X is $\log_\mu 2$.   □

Exercise 8.23   The Sierpinski carpet fractal is constructed as follows: The unit square is divided into nine equal boxes, and the open central box is deleted. This process is repeated for each of the remaining sub-boxes, and infimum.

  (i) Find the IFS of similitudes which generates the Sierpinski carpet;

  (ii) Find the Hausdorff dimension of the limit set;

  (iii) Show that the Sierpinski carpet has zero area.   □

Exercise 8.24   For the classical Cantor set C, show that

  (i) C contains no interval of positive length.

  (ii) C has no isolated points, that is, if $x \in C$, then every neighborhood of x will intersects C in more than one point.   □

Exercise 8.25   Show that the classical Cantor set C contains an uncountable number of points.□

Exercise 8.26   If you instead of the classical Cantor set proceed as suggested before, and remove the 2nd and 4th fourth, i.e., in step one remove $(\frac{1}{4}, \frac{1}{2})$, $(\frac{3}{4}, 1]$, etc., you arrive at a Cantor set C'. Find the IFS that generates the set C' and determine its Hausdorff dimension.   □

Exercise 8.27   The Menger sponge M is constructed starting from a cube with sides 1. In the first step, we divide it into 27 similar cubes and remove the center cube plus all cubes at the center of the faces. Then we continue to do the same to each of the 27 new cubes. Determine the Hausdorff dimension of M.   □

NOTES FOR CHAPTER 8

The term *fractal* is coined by the mathematician Benoît Mandelbrot in 1975, derived by him from the Latin word "fractus," which means "broken" or "fracture". A fractal is a geometric object with a *self-similar* property studied in this chapter. The earliest ideas of fractals can be dated back to Karl Weierstrass, Helge von Koch, Georg Cantor, and Felix Hausdorff, among others. Some of the most famous fractals are named after Cantor, Sierpinski, Peano, Koch, Harter-Heighway, Menger, Julia, etc. The major technique we note here is the IFS (iterated function systems) in Section 8.3 to define the fractals.

But having self-similarity alone is not sufficient for an object to be a fractal. For example, a straight line contains copies of itself at finer and finer scales. But it does not qualify as a fractal, as a straight line has the same Hausdorff dimension as the topological dimension, which is one. Section 8.2 gives a concise account of the Hausdorff measure and dimension, which will also be of major utility in Chapter 8.

CHAPTER 9

Rapid Fluctuations of Chaotic Maps on $\mathbb{R}^N$

We have studied in Chapter 2 the use of total variations to characterize the chaotic behavior of interval maps. In this chapter, we will generalize such an approach to maps on multi-dimensional spaces.

First, we note that for a chaotic map on a multi-dimensional space, chaos may happen only on a lower-dimensional manifold C, on the complement of which the map's behavior can be quite orderly. The subset C generally is expected to have a fractal structure and a fractional dimensionality. Thus, to characterize chaotic properties in multi-dimensional spaces, we must rely on the use of the Hausdorff dimensions, the Hausdorff measure and fractals developed in Chapter 8.

9.1 TOTAL VARIATION FOR VECTOR-VALUE MAPS

Let $\mathrm{Lip}(\mathbb{R}^N)$ denote the class of all Lipschitz continuous maps from $\mathbb{R}^N$ to $\mathbb{R}^N$.

Definition 9.1 Let $f \in \mathrm{Lip}(\mathbb{R}^N)$ and A be an s-set of $\mathbb{R}^N$. The *total variation* of f on A (with respect to the s-Hausdorff dimension) is defined by

$$\mathrm{Var}_A^{(s)}(f) = \sup \left\{ \sum_{i=1}^n \mathcal{H}^s(f(C_i)) \mid n \in \mathbb{N} = \{1, 2, 3, \ldots\}, \right.$$
$$\left. A \supset \bigcup_{i=1}^n C_i, \, C_i \cap C_j = \emptyset, i \neq j \right\}.$$

Since the total variation in the definition is always bounded in the case that $f \in \mathrm{Lip}(\mathbb{R}^N)$. $\square$

The following are equivalent definitions.

Lemma 9.2 Let $f \in \mathrm{Lip}(\mathbb{R}^N)$ and A be an s-set of $\mathbb{R}^N$. Then

$$\mathrm{Var}_A^{(s)}(f) = \sup \left\{ \sum_{i=1}^n \mathcal{H}^s(f(C_i)) \mid \forall n \geq 1, \right.$$
$$\left. A \supset \bigcup_{i=1}^n C_i, \, C_i \text{ are } s\text{-sets}, C_i \cap C_j = \emptyset, i \neq j \right\} \tag{9.1}$$

$$\mathrm{Var}_A^{(s)}(f) = \sup\left\{\sum_{i=1}^n \mathcal{H}^s(f(C_i)) \mid \forall n \geq 1,\right.$$
$$\left. A \supset \bigcup_{i=1}^n C_i, \; C_i \text{ are } s\text{-sets}, \; \mathcal{H}^s(C_i \cap C_j) = 0, i \neq j\right\}. \tag{9.2}$$

Proof. Denote by V_1 and V_2, respectively, the right-hand sides of (9.1) and (9.2). It suffices to show that

$$\mathrm{Var}_A^{(s)}(f) = V_1, \qquad V_1 = V_2. \tag{9.3}$$

From (9.1), it follows $V_1 \leq \mathrm{Var}_A^{(s)}(f)$. The converse inequality follows from the fact that a subset C of an s-set is not an s-set itself if and only if $\mathcal{H}^s(C) = 0$, which implies $\mathcal{H}^s(f(C)) = 0$ since f is Lipschitz continuous. Thus,

$$\mathrm{Var}_A^{(s)}(f) = V_1.$$

We now prove the second equality in (9.3). It is obvious that $V_1 \leq V_2$. Conversely, let $C_i, i = 1, \ldots, n$ be s-sets such that

$$A \supset \bigcup_{i=1}^n C_i, C_i \text{ are } s-\text{sets}, \qquad \mathcal{H}^s(C_i \cap C_j) = 0, i \neq j.$$

Define $C_i', i = 1, \ldots, n$, inductively by

$$C_1' = C_1, C_i' = C_i - \bigcup_{j=1}^{i-1} C_j' = C_i - C_i \cap \bigcup_{j=1}^{i-1} C_j', \quad i = 2, \ldots n.$$

Then $C_i', i = 1, \ldots n$ are disjoint and

$$\bigcup_{i=1}^n C_i' = \bigcup_{i=1}^n C_i.$$

Since

$$C_i \cap \bigcup_{j=1}^{i-1} C_j' \subset C_i \cap \bigcup_{j=1}^{i-1} C_j = \bigcup_{j=1}^{i-1} (C_i \cap C_j),$$

we have

$$\mathcal{H}^s\left(C_i \cap \bigcup_{j=1}^{i-1} C_j'\right) \leq \sum_{j=1}^{i-1} \mathcal{H}^s\left(C_i \cap C_j\right) = 0. \tag{9.4}$$

Finally, for $i = 1, \ldots, n$, we have

$$C_i' \subset C_i = C_i' \bigcup \left(C_i \cap \bigcup_{j=1}^{i-1} C_j' \right),$$

which implies

$$f(C_i') \subset f(C_i) = f(C_i') \cup f\left(C_i \cap \bigcup_{j=1}^{i-1} C_j' \right).$$

Thus,

$$\mathcal{H}^s(f(C_i')) \le \mathcal{H}^s(f(C_i)) \le \mathcal{H}^s(f(C_i')) + \mathcal{H}^s\left(f(C_i \cap \bigcup_{j=1}^{i-1} C_j') \right) = \mathcal{H}^s(f(C_i')),$$

since

$$\mathcal{H}^s\left(f(C_i \cap \bigcup_{j=1}^{i-1} C_j') \right) = 0$$

by (9.4) and the Lipschitz property of f.

Summing up the above, we have

$$\sum_{i=1}^n \mathcal{H}^s(f(C_i)) = \sum_{i=1}^n \mathcal{H}^s(f(C_i')).$$

Thus, $V_2 \le V_1$. So $V_1 = V_2$.   □

We now state some properties on the total variation of f.

Lemma 9.3   Let $f \in \mathrm{Lip}(\mathbb{R}^N)$ and A be an s-set of $\mathbb{R}^N$. Then

(i) $\mathrm{Var}_A^{(s)}(f) \le (\mathrm{Lip}(f))^s \mathcal{H}^s(A)$;

(ii) for each $\lambda > 0$, we have

$$\mathrm{Var}_A^{(s)}(\lambda f) = \lambda^s \, \mathrm{Var}_A^{(s)}(f);$$

(iii) if $f \in \mathrm{Lip}(\mathbb{R}^1)$ and $A = [a, b]$ is a bounded interval, then

$$\mathrm{Var}_A^{(1)}(f) = V_{[a,b]}(f),$$

where $V_{[a,b]}(f)$ is the usual total variation of f on $[a, b]$.

Proof. (i) and (ii) follows from the definition and parts (v) and (vi), respectively, in Lemma 8.5. For (iii), by Lemma 8.6, we have

$$\text{Var}_{[a,b]}^{(1)}(f) = \sup\left\{\sum_{i=1}^{n}\mathcal{H}^1(f[x_{i-1},x_i]) \mid \forall n \geq 1 \ x_0 = a < x_1 < \cdots < x_n = b\right\}$$

$$= \sup\left\{\sum_{i=1}^{n}\mathcal{L}^1(f[x_{i-1},x_i]) \mid \forall n \geq 1 \ x_0 = a < x_1 < \cdots < x_n = b\right\}.$$

So

$$V_{[a,b]}(f) \leq \text{Var}_{[a,b]}^{(1)}(f).$$

Conversely, for any $\varepsilon > 0$, there exist a positive integer n_0 and a partition of $[a,b]$: $x_0 = a < x_1 < \cdots < x_{n_0} = b$ such that

$$\text{Var}_{[a,b]}^{(1)}(f) \leq \sum_{i=1}^{n_0}\mathcal{L}^1(f[x_{i-1},x_i]) + \varepsilon.$$

For $i = 1, \ldots n_0$, let

$$f(\xi_i) = \max_{x_{i-1}\leq x\leq x_i}\{f(x)\}, \quad x_{i-1} \leq \xi_i \leq x_i,$$

$$f(\eta_i) = \min_{x_{i-1}\leq x\leq x_i}\{f(x)\}, \quad x_{i-1} \leq \eta_i \leq x_i.$$

Since f is continuous, f can attain its maximum and minimum on any bounded closed interval. Thus,

$$\mathcal{L}^1(f([x_{i-1},x_i])) = |f(\xi_i) - f(\eta_i)|,$$

where $\mathcal{L}^1$ is the 1-dimensional Lebesgue measure.

By adding the points ξ_i, η_i into the partition if necessary, we have a new partition: $x_0' = a < x_1' < \cdots x_{n'}' = b$ $(n' \geq n_0)$, in which ξ_i, η_i are the partition points. Thus, by definition

$$V_{[a,b]}(f) + \varepsilon \geq \sum_{i=1}^{n'}|f(x_i') - f(x_{i-1}')| + \varepsilon$$

$$\geq \sum_{i=1}^{n_0}|f(\xi_i) - f(\eta_i)| + \varepsilon$$

$$\geq \text{Var}_{[a,b]}^{(1)}(f).$$

Hence, $\text{Var}_{[a,b]}^{(1)}(f) \leq V_{[a,b]}(f)$ by the arbitrariness of ε and $\text{Var}_{[a,b]}^{(1)}(f) = V_{[a,b]}(f)$. $\qquad\square$

Property (iii) in Lemma 9.3 shows that Definition 9.1 is indeed a generalization of bounded variation of one-dimensional maps to vector-valued maps.

9.2   RAPID FLUCTUATIONS OF MAPS ON $\mathbb{R}^N$

Similar to Definition 2.7 for one-dimensional dynamical systems, we now introduce a new notion to describe the complexity of multi-dimensional dynamical systems.

Definition 9.4   Let $D \subset \mathbb{R}^N$ and $f \colon D \to D$ be Lipschitz continuous. If there exists an s-set $A \subset D$ ($0 \leq s \leq N$) such that $\mathrm{Var}_A^{(s)}(f^n)$ on A grows exponentially as $n \to \infty$, then we say that f has rapid fluctuations of dimension s.   □

We remark that the exact computation for the total variation $\mathrm{Var}_A^{(s)}(f^n)$ is nearly impossible, since, in general, we can not calculate the exact value of Hausdorff measures for s-sets. Fortunately, only the estimates of the lower bound of $\mathrm{Var}_A^{(s)}(f^n)$ are needed if we want to show f has rapid fluctuations.

For a Lipschitz continuous map f, f may have rapid fluctuations of different dimensions. The supremum of such dimensionality numbers is called *the dimension of rapid fluctuations*.

Example 9.5   Consider the tent map $f_\mu \colon [0, 1] \to \mathbb{R}$ defined by

$$f_\mu(x) = \begin{cases} \mu x, & 0 \leq x \leq 1/2, \\ \mu(1 - x), & 1/2 < x \leq 1, \end{cases} \qquad (\mu > 0),$$

cf. the special case in Example 4.4 and Fig. 5.1. For $\mu > 2$, let

$$X = \{x \in [0, 1] \mid f_\mu^n(x) \in [0, 1], \quad \text{for all} \quad n = 0, 1, 2, \ldots\}.$$

X is the largest invariant set of f_μ contained in $[0, 1]$. It is an extant result that f_μ is chaotic on X. In fact, X is also a self-similar set generated by the IFS of similitudes

$$S_0(x) = \frac{1}{\mu}x, \qquad S_1(x) = \frac{1}{\mu}x + 1 - \frac{1}{\mu}.$$

Since $\mu > 2$, the open set condition holds by taking

$$V = (0, 1).$$

The set X is self-similar and has a Hausdorff dimension of $\ln 2 / \ln \mu$ by Proposition 8.15. Therefore, we conclude that for $\mu > 2$, f_μ has rapid fluctuations on an s-set X with $s = \ln 2 / \ln \mu$.

In fact, for any positive integer n, there exists 2^n subintervals $J_i^n = [a_i^n, a_{i+1}^n]$, $i = 0, 1, \ldots, 2^n - 1$ with

$$a_0^n = 0, \qquad a_i^n < a_{i+1}^n, \qquad a_{2^n}^n = 1,$$

such that f_μ^n is strictly monotone on J_i^n and $f_\mu^n(J_i^n) = [0, 1]$, $i = 0, 1, \ldots, 2^n - 1$. Therefore,

$$\mathrm{Var}_X^{(s)}(f_\mu^n) \geq \sum_{i=0}^{2^n-1} \mathcal{H}^s(f_\mu^n(X \cap J_i^n)) = 2^n \mathcal{H}^s(X). \qquad \square$$

A more interesting example is the standard Smale horseshoe map.

Example 9.6   Consider the standard Smale horseshoe map φ introduced in Section 7.1 with $\lambda = 3$. We know that from Theorem 6.52 that φ is chaotic on the invariant set Λ. But now we consider the rapid fluctuations property of φ.

First, since φ is a homeomorphism, it has no rapid fluctuations of dimension 2. But it has rapid fluctuations of dimension 1, since each vertical line in the square Q defined in (7.2) is split into two vertical lines under φ.

Next, we show that for each $s \in (0, (\ln 6/\ln 3) - 1)$, φ has rapid fluctuations of dimension $1 + s$, which illustrates that φ has *higher dimensional chaos*! In fact, for any $s \in (0, (\ln 6/\ln 3) - 1)$, taking an arbitrary s-set X in $[-1, 1]$ and letting $A = X \times [-1, 1]$, we have $A \subset Q$ as an s-set and

$$\dim_{\mathcal{H}}(A) = \dim_{\mathcal{H}}(X) + \dim_{\mathcal{H}}([-1, 1]) = s + 1,$$

by Proposition 8.16. To prove our claim, it suffices to show that $\mathrm{Var}_A^{(s+1)}(f^n)$ grows exponentially as $n \to \infty$.

By the definition of the map φ, the image of A under φ contains six smaller "copies" of A under similitude mappings with a contraction ratio $\frac{1}{3}$. Therefore,

$$\mathcal{H}^{1+s}(f(A)) \geq 6\left(\frac{1}{3}\right)^{1+s} \mathcal{H}^{1+s}(A) = (1 + \delta)\mathcal{H}^{1+s}(A),$$

where

$$\delta = 6\left(\frac{1}{3}\right)^{1+s} - 1$$

is a positive constant which depends only on s. Inductively, we have

$$\mathcal{H}^{1+s}(f^n(A)) \geq (1 + \delta)^n \mathcal{H}^{1+s}(A).$$

Thus, $\mathrm{Var}_A^{(s+1)}(f^n) \geq (1 + \delta)^n \mathcal{H}^{1+s}(A)$ grows exponentially as $n \to \infty$. $\qquad \square$

We show next that rapid fluctuations are unchanged under Lipschitz conjugacy.

Definition 9.7   Let $D_1 \subset \mathbb{R}^{N_1}$, $D_2 \subset \mathbb{R}^{N_2}$, $f: D_1 \to D_1$ and $g: D_2 \to D_2$ be Lipschitz continuous maps. We say that f and g are Lipschitz conjugate if there exists a bi-Lipschitz map h from D_1 to D_2 such that

$$h \circ f = g \circ h. \qquad (9.5)$$

□

Lemma 9.8   Let f and g have Lipschitz conjugacy h. Then

(i) f has rapid fluctuations of dimension s if and only if g has rapid fluctuations of dimension s.

(ii) Let A be a s-set in D_1 for some $s \in (0, N]$. Then $\mathrm{Var}_A^{(s)}(f^n)$ grows unbounded if and only if $\mathrm{Var}_{h(A)}^{(s)}(g^n)$ grows unbounded as $n \to \infty$, where the bi-Lipschitz map $h \colon D_1 \to D_2$ is given by (9.5).

Proof. (i) Let h be bi-Lipschitz continuous such that $h \circ f = g \circ h$. Then $f = h^{-1} \circ g \circ h$. Let A be an s-set. Then $h(A)$ is also an s-set by part (v) in Lemma 8.5. For any $k > 1$, subsets $C_1, C_2, \ldots, C_k$ satisfy $C_i \cap C_j = \emptyset$, $i \neq j$, and $A \supset \bigcup_{i=1}^{k} C_i$ if and only if compact subsets $h(C_1), h(C_2), \ldots, h(C_k)$ satisfy the very same property with $h(A) \supset \bigcup_{i=1}^{k} h(C_i)$. Thus, by the definition of conjugacy and by part (v) of Lemma 8.5, we have

$$\mathrm{Var}_A^{(s)}(f^n) \leq (\mathrm{Lip}(h^{-1}))^s \, \mathrm{Var}_{h(A)}^{(s)}(g^n).$$

Therefore, if there is an s-set A such that $\mathrm{Var}_A(f^n)$ grows exponentially as $n \to \infty$, then so does $\mathrm{Var}_{h(A)}(g^n)$, and vice versa.

The proof of part (ii) is similar, so we omit it.   □

9.3   RAPID FLUCTUATIONS OF SYSTEMS WITH QUASI-SHIFT INVARIANT SETS

From Chapter 6.6, it is known that a compact dynamical system has complex behavior if it has a quasi shift invariant set. In the section, we show that f has rapid fluctuations provided that f has a quasi shift invariant set.

Theorem 9.9   Let (X, f) be a compact dynamical system and f be Lipschitz continuous. Assume that assumption (i) in Theorem 6.53 holds. That is, there there exist k subsets $A_0, A_1, \ldots, A_{k-1}$ of X with $k \geq 2$ which are mutually disjoint such that

$$f(A_i) \supset \bigcup_{j=0}^{k-1} A_j, \qquad i = 0, 1, \ldots, k-1. \tag{9.6}$$

If there exists an $i_0 \in \{0, 1, \ldots k - 1\}$ such that A_{i_0} is an s-set, then A_i is also an s-set and $\mathrm{Var}_{A_i}^{(s)}(f^n)$ grows exponentially for every $i \in \{0, 1, \ldots k - 1\}$. Consequently, f has rapid fluctuations of dimension s.   □

It follows from Corollary 6.54 that condition (9.6) implies that f has a quasi-shift invariant set.

Proof of Theorem 9.9. Denote by L the Lipschitz constant of f. By assumption, for any $i = 0, 1, \ldots k - 1$, and Lemma 8.5, we have

$$0 < \mathcal{H}^s(A_{i_0}) \le \mathcal{H}^s \left(\bigcup_{j=0}^{k-1} A_j \right) \le \mathcal{H}^s(f(A_i)) \le L^s \mathcal{H}^s(A_i)$$

$$\le L^s \mathcal{H}^s \left(\bigcup_{j=0}^{k-1} A_j \right) \le L^s \mathcal{H}^s(f(A_{i_0})) \le L^{2s} \mathcal{H}^s(f(A_{i_0})) < \infty.$$

Since A_{i_0} is an s-set, so is A_i.

Now we prove that f has rapid fluctuations. Denote

$$\delta = \min \left\{ \mathcal{H}^s(A_i), \qquad i = 0, 1, \ldots, k - 1 \right\}.$$

Then $\delta > 0$ since A_i $(i = 0, 1, \ldots, k - 1)$ is an s-set. For a given i, let

$$J_{ij} = f^{-1}(A_j) \cap A_i, \qquad j = 0, 1, \ldots, k - 1.$$

Then J_{ij} $(j = 0, 1, \ldots, k - 1)$ are mutually disjoint and

$$f(J_{ij}) = A_j. \tag{9.7}$$

Thus, by definition, we have

$$\mathrm{Var}_{A_i}^{(s)}(f) \ge \sum_{j=0}^{k-1} \mathcal{H}^s(f(J_{ij})) = \sum_{j=0}^{k-1} \mathcal{H}^s(A_j) \ge k\delta.$$

On the other hand, from (9.7) and the given assumptions, it follows that

$$f^2(J_{ij}) \supset \bigcup_{\ell=1}^{k-1} A_\ell, \qquad j = 0, 1, \ldots, k - 1.$$

In the same way, we can find subsets of $J_{ij\ell}$ of J_{ij} which are mutually disjoint such that

$$f^2(J_{ij\ell}) = A_\ell, \qquad j, \ell = 0, 1, \ldots, k - 1.$$

Thus,

$$\mathrm{Var}_{A_i}^{(s)}(f^2) \ge \sum_{j,\ell=0}^{k-1} \mathcal{H}^s(f^2(J_{ij\ell}))$$

$$= \sum_{j,\ell=0}^{k-1} \mathcal{H}^s(A_\ell) \ge k^2 \delta.$$

Repeating the above procedures, we can prove by induction that

$$\text{Var}(A_i, f^n) \geq k^n \delta.$$

So $\text{Var}_{A_i}^{(s)}(f^n)$ grows exponentially. This completes the proof. □

We know from Theorem 6.58 that for a C^1 map $f: \mathbb{R}^N \longrightarrow \mathbb{R}^N$ with a snap-back repeller p, there exists a positive integer r such that f^r has a shift invariant set of order 2 with respect to the one-sided shift σ^+. Thus, we have the following.

Corollary 9.10 Let $f: \mathbb{R}^N \longrightarrow \mathbb{R}^N$ be C^1. If f has a snap-back repeller p, then f has rapid fluctuations of dimension N.

Proof. From the proof of Theorem 6.58, there exist two closed neighborhoods A_1, A_2 of p and a positive integer r such that

$$A_1 \cap A_2 = \emptyset, \qquad f^r(A_1) \cap f^r(A_2) \supset A_1 \cup A_2.$$

Thus, A_i $(i = 1, 2)$ are N-sets. By Theorem 6.58, $\text{Var}_{A_i}^{(N)}((f^r)^n)$ grow exponentially as $n \to \infty$.

On the other hand, since f is C^1, f is locally Lipschitz. Denote by L the Lipschitz constant on the neighborhood of p which contains both A_1 and A_2. For any positive integer $m > r$, let $m = rn - \ell$ with $0 \leq \ell < r$. Then we have

$$\text{Var}_{A_i}^{(N)}((f^r)^n) \leq L^{\ell N} \text{Var}_{A_i}^{(N)}(f^m).$$

Thus, $\text{Var}_{A_i}^{(N)}(f^m)$ grow exponentially as $n \to \infty$. That is, f has rapid fluctuations of dimension N. □

9.4   RAPID FLUCTUATIONS OF SYSTEMS CONTAINING TOPOLOGICAL HORSESHOES

In Chapter 7, we have discussed that the standard Smale horseshoe and its general case have shift invariant sets with respect to the two-sided shift σ. So they have complex dynamical behavior. In this section, we discuss the more general dynamical systems of topological horseshoes which include the Smale horseshoe as a special case. We show that such a map with topological horseshoe has rapid fluctuations.

Definition 9.11   Assume that X is a separable metric space and consider the dynamical system (X, f). If there is a locally connected and compact subset Q of X such that

  (i) the map $f: Q \to X$ is continuous;

(ii) there are two disjoint and compact sets Q_0 and Q_1 of Q such that each component of Q intersects both Q_0 and Q_1;

(iii) Q has a crossing number $M \geq 2$ (see below),

then (X, f) is said to have a topological horseshoe.   □

From the above, we define a *connection* Γ between Q_0 and Q_1 as a compact connected subset of Q that intersects both Q_0 and Q_1. A *preconnection* γ is defined as a compact connected subset of Q for which $f(\gamma)$ is a connection. We define the *crossing number* M to be the largest number such that every connection contains at least M mutually disjoint preconnections.

Example 9.12   Let φ be the standard Smale horseshoe map defined in Section 4.1. Then φ satisfies the topological horseshoe assumption on $V = Q \cap \varphi(Q)$ with crossing number $M = 2$ and

$$Q_0 = V \cap \{y = -1\}, \qquad Q_1 = V \cap \{y = 1\},$$

where $Q = [-1, 1] \times [-1, 1]$.   □

Example 9.13   Let ψ be the general Smale horseshoe map defined in Section 7.2. That is, $\psi : Q \to \mathbb{R}^2$ satisfies assumptions (A1) and (A2) in Section 7.2, where $Q = [-1, 1] \times [-1, 1]$. Let

$$V = \bigcup_{j=0}^{N-1} V_j.$$

Then ψ satisfies the topological horseshoe assumption on V with the crossing number $M = N$. The proof of this is left as an exercise.   □

Kennedy and Yorke [45] studied the chaotic properties of dynamical systems with topological horseshoes. We state one of the main theorems therein, below.

Lemma 9.14   ([45, Theorem 1]) Assume that f has a topological horseshoe. Then there is a closed invariant set $Q_I \subset Q$ for which $f|_{Q_I} : Q_I \to Q_I$ is semiconjugate to a one-sided M-shift.   □

We now discuss rapid fluctuations of dynamical systems with a topological horseshoe. We prove that such kind of systems has rapid fluctuations of dimension at least 1. To this end, we first need the following.

Lemma 9.15   If Γ is a connection of Q according to Definition 9.11, then the Hausdorff dimension of Γ is at least 1.

Proof. Let $x_0 \in \Gamma$ be fixed. Define $h \colon \Gamma \to \mathbb{R}^+$ by

$$h(x) = d(x_0, x),$$

where $d(\cdot, \cdot)$ is the distance in X.

Let $a = \max\limits_{x \in \Gamma} h(x)$. Then $h \colon \Gamma \to [0, a]$ is a non-expanding and onto map by the connectedness of Γ. Thus,

$$0 < \mathcal{H}^1([0, a]) = \mathcal{H}^1(h(\Gamma)) \leq \mathcal{H}^1(\Gamma).$$

This implies that $\dim_{\mathcal{H}}(\Gamma) \geq 1$. $\qquad\square$

Assume that f has topological horseshoes. Denote by $\mathcal{S}$ the set of all connections of Q and

$$s = \inf\{\dim_{\mathcal{H}}(\Gamma), \ \Gamma \in \mathcal{S}\}. \tag{9.8}$$

We have $s \geq 1$ from Lemma 9.15.

Theorem 9.16 Assume that f has a topological horseshoe and is Lipschitz continuous. Let s be defined as in (9.8). If there exists a $\Gamma_0 \in \mathcal{S}$ such that $\dim_{\mathcal{H}}(\Gamma_0) = s$ and

$$\mathcal{H}^s(\Gamma_0) = \inf\left\{\mathcal{H}^s(\Gamma) \mid \Gamma \in \mathcal{S} \quad \text{and} \quad \dim_{\mathcal{H}}(\Gamma) = s\right\} > 0, \tag{9.9}$$

then $\mathrm{Var}_{\Gamma_0}(f^n)$ grows exponentially as $n \to \infty$. Thus, f has rapid fluctuations of dimension s.

Proof. Since $\Gamma_0 \in \mathcal{S}$, by assumptions Γ_0 has at least M mutually disjoint preconnections, which are denoted by $\Gamma_{01}, \Gamma_{02}, \dots, \Gamma_{0M}$. That is $\Gamma_{0i} \subset \Gamma_0$ and $f(\Gamma_{0i}) \in \mathcal{S}$, for $i = 1, 2, \dots, M$. Hence, we have

$$\mathcal{H}^s(f(\Gamma_{0i})) \leq L\mathcal{H}^s(\Gamma_{0i}) \leq L\mathcal{H}^s(\Gamma_0),$$

where L is the Lipschitz constant. It follows from (9.9) that $\dim_{\mathcal{H}}(f(\Gamma_{0i})) = s$ and

$$\mathrm{Var}_{\Gamma_0}(f) \geq \sum_{i=0}^{M} \mathcal{H}^s(f(\Gamma_{0i})) \geq M\mathcal{H}^s(\Gamma_0).$$

For each i $(1 \leq i \leq M)$, since $f(\Gamma_{0i}) \in \mathcal{S}$, again $f(\Gamma_{0i})$ has at least M mutually disjoint preconnections, denoted by $\gamma_{0i}^1, \dots, \gamma_{0i}^M$. So

$$f(\gamma_{0i}^1) \in \mathcal{S}, \qquad \bigcup_{\ell=1}^{M} \gamma_{0i}^\ell \subset f(\Gamma_{0i}).$$

The latter implies that there exist M mutually disjoint subsets $\Gamma_{0i}^1, \dots, \Gamma_{0i}^M$ of Γ_{0i} such that $\gamma_{0i}^\ell = f(\Gamma_{0i}^\ell)$. We thus obtain M^2 mutually disjoint subsets Γ_{0i}^ℓ, $i, \ell = 1, \dots, M$ of Γ_0. Therefore,

$$\mathrm{Var}_{\Gamma_0}(f^2) \geq \sum_{i,\ell=1}^{M} \mathcal{H}^s(f^2(\Gamma_{0i}^\ell)) = \sum_{i,\ell=1}^{M} \mathcal{H}^s(f(\gamma_{0i}^\ell)) \geq M^2\mathcal{H}^s(\Gamma_0).$$

Repeating the above procedure, we can prove by induction that

$$\mathrm{Var}_{\Gamma_0}(f^n) \geq M^n \mathcal{H}^s(\Gamma_0).$$

That is, $\mathrm{Var}_{\Gamma_0}(f^n)$ grows exponentially as $n \to \infty$. Thus, f has rapid fluctuations of dimension s.
□

9.5   EXAMPLES OF APPLICATIONS OF RAPID FLUCTUATIONS

In this section, we give two examples to illustrate rapid fluctuations: one from nonlinear economic dynamics, and the other from a predator-prey model.

Example 9.17   Benhabib and Day [4, 5] used two types of one-dimensional systems as models for dynamic consumer behavior. One is the well-known logistic type. The second is an exponential type. Dohtani [22] presented two classes of examples of the Benhabib–Day model in a multi-dimensional space, which are governed by the Lotka–Volterra type

$$x_i(n+1) = f_i(x(n)), \qquad f_i(x_1, \ldots, x_N) = x_i \left(a - \sum_{j=1}^{N} b_{ij} x_j \right), \qquad (9.10)$$

or by the exponential type

$$x_i(n+1) = g_i(x(n)), \qquad g_i(x_1, \ldots, x_N) = x_i \exp \left(a - \sum_{j=1}^{N} b_{ij} x_j \right), \qquad (9.11)$$

where $i \in H = \{1, 2, \ldots, N\}$. The variable x_i ($i \in H$) is the amount of the i-th good consumed within a given period, and the constants $a > 0$, b_{ij} are the parameters with respect to the economic environment. See [22] for details. The chaotic behavior of (9.10) and (9.11) is proved in Theorem 9.18.
□

Theorem 9.18   Let

$$Q = (1, \ldots, 1)^T \in \mathbb{R}^N, \qquad B = [b_{ij}] \in \mathbb{R}^{N \times N}.$$

Suppose that the matrix B is nonsingular and each entry in $B^{-1}Q$ is positive.

(1) If $1 < a < 4$ and the logistic map

$$\alpha(x) = ax(1-x),$$

from $[0, 1]$ to itself, has a periodic orbit whose period is not a power of 2, then the Lotka–Volterra system (9.10) has rapid fluctuations of dimension 1.

(2) If the exponential map

$$\beta(x) = x \exp(a - x), \tag{9.12}$$

from $\mathbb{R}_+$ to itself, has a periodic orbit whose period is not a power of 2, then the exponential system (9.11) has rapid fluctuations of dimension 1. □

Diamond [21] introduced the following definition.

Definition 9.19 (Chaos in the sense of P. Diamond) Let $D \subset \mathbb{R}^N$ and $f: D \to D$ be continuous. We said that f is chaotic in the sense of Diamond if

(i) for every $n = 1, 2, 3, \ldots$, there is an n-periodic orbit in D;

(ii) there is an uncountable set S of D, which contains no periodic points and satisfies

    (a) $f(S) \subset S$;

    (b)

$$\limsup_{n \to \infty} |f^n(p) - f^n(q)| > 0, \qquad \forall p, q \in S, \ p \neq q,$$
$$\liminf_{n \to \infty} |f^n(p) - f^n(q)| = 0, \qquad \forall p, q \in S;$$

    (c) for every p in S and every periodic point q in D;

$$\limsup_{n \to \infty} |f^n(p) - f^n(q)| > 0. \qquad\qquad □$$

Remark 9.20 It is known that the logistic map $\alpha(x)$ in Theorem 9.18 above has a periodic point whose period is not a power of 2 if $a > a^* = 3.59 \cdots$. Since the exponential map $\beta(x)$ is strictly increasing in $[0, 1]$ and strictly decreasing in $[1, \infty]$, respectively, and $\beta(x) \to 0$ as $x \to +\infty$, $\beta(x)$ has a period-three point if $\beta^3(1) < 1$. A sufficient condition for this is $a \geq 3.13$ ([22]). If $\beta^3(1) \leq a$, then $\beta(x)$ has a period-6 point. This is the case when $a \geq 2.888$.

Dohtani [22] has established that the Lotka–Volterra system (9.10) generates chaos in the sense of Diamond if $a > 3.84$, and the exponential type system (9.11) is chaotic in the same sense if $a > 3.13$. □

Proof of Theorem 9.18. Let $W = aB^{-1}Q = (w_1, w_2, \ldots, w_N)^T$. Then $W > 0$ by the assumptions. Denote

$$\Omega = \{rW \mid 0 \leq r \leq 1\}.$$

Then Ω is the line segment in $\mathbb{R}^N$ that connects the origin with W, so it is a 1-set. For any $x \in \Omega$, there exists a positive constant $r \in [0, 1]$ such that $x = rW$. Then, by (9.10), we have

$$
\begin{aligned}
f(x) &\equiv (f_1(x), \ f_2(x), \dots, \ f_N(x))^T \\
&= r \ \mathrm{diag}(w_1, \ w_2, \dots, \ w_N)(aQ - rBW) = r(1 - r)aW,
\end{aligned}
\tag{9.13}
$$

where $\mathrm{diag}(\dots)$ is a diagonal matrix with the indicated diagonal entries. Noting that $r(1 - r) \leq \frac{1}{4}$ for any $0 \leq r \leq 1$, we have $0 \leq r(1 - r)a \leq 1$ for $1 < a < 4$. It follows from (9.13) that $f(x) \in \Omega$. This means that Ω is an invariant set under f.

   If the logistic map $\alpha(x)$ has a periodic point whose period is not a power of 2, then there exists a positive integer k such that $\alpha^k(\cdot)$ is *strictly turbulent*, i.e., there are compact intervals $J, K \subset [0, 1]$ with $J \cap K = \emptyset$ and

$$
\alpha^k(J) \cap \alpha^k(K) \supset J \cup K.
$$

Let

$$
J' = JW \equiv \{rW \mid r \in J\}, \qquad K' = KW \equiv \{rW \mid r \in K\}.
$$

Then J' and K' are two compact 1-sets in $\mathbb{R}^N$ with empty intersection and

$$
f^k(J') \cap f^k(K') \supset J' \cup K'.
$$

Thus, system (9.13) has rapid fluctuations of dimension 1 by Theorem 9.16.

   The corresponding property for the exponential type system (9.11) can be proved similarly.

□

Example 9.21   We consider the following predator-prey model

$$
\begin{cases}
\dfrac{dx}{dt} = x(t)[\mu_1 - \mu_1 x(t) - s_1 y(t)], \\[2mm]
\dfrac{dy}{dt} = y(t)[-\mu_2 + s_2 y(t)].
\end{cases}
\tag{9.14}
$$

Let $t_n = nh$ where h is a step size. Applying the variation of parameter formula to each equation in (9.14), one obtains

$$
\begin{cases}
x(t_{n+1}) = x(t_n) \exp \left(\displaystyle\int_{t_n}^{t_{n+1}} [\mu_1 - \mu_1 x(\xi) - s_1 y(\xi)] d\xi \right) \\[1mm]
\qquad\quad\ \approx x(t_n) \exp(\mu_1 - \mu_1 x(t_n) - s_1 y(t_n)]h), \\[2mm]
y(t_{n+1}) = y(t_n) \exp \left(\displaystyle\int_{t_n}^{t_{n+1}} [-\mu_2 + s_2 y(\xi)] d\xi \right) \\[1mm]
\qquad\quad\ \approx y(t_n) \exp([-\mu_2 + s_2 y(t_n)]h).
\end{cases}
\tag{9.15}
$$

Thus, the following difference equation system gives an approximate numerical scheme for (9.15):

$$\begin{cases} x(t_{n+1}) = x(t_n) \exp(\mu_1 - \mu_1 x(t_n) - s_1 y(t_n)]h), \\ y(t_{n+1}) = y(t_n) \exp([-\mu_2 + s_2 y(t_n)]h). \end{cases} \tag{9.16}$$

Denoting $x(t_n) = x_n$ and $y(t_n) = y_n$, re-scaling $hs_1 y_n \to y_n$ and also re-scaling parameters $h\mu_1 \to \mu_1$ and $hs_2 \to h$, we rewrite (9.16) as

$$\begin{cases} x_{n+1} = x_n \exp[\mu_1 - \mu_1 x_n - y_n], \\ y_{n+1} = y_n \exp[-\mu_2 + sx_n], \end{cases} \tag{9.17}$$

where $s = s_2$. We let $F = F_{\mu_1, \mu_2, s} : \mathbb{R}^2 \to \mathbb{R}^2$ be the map signifying the right hand side of (9.17), i.e.,

$$F(x, y) = (x \exp[\mu_1 - \mu_1 x - y], \quad y \exp[-\mu_2 + sx]).$$

Now we consider the dynamical system $(\mathbb{R}^2, F)$. We show that F has rapid fluctuations of dimension 1.

Denote

$$h(x) = xe^{\mu_1 - \mu_1 x}, \tag{9.18}$$
$$g_1(x, y) = \mu_1 h(x)(1 - e^{-y}) - y(1 + e^{-\mu_2 + sx}),$$
$$g_2(x, y) = -2\mu_2 + s(x + h(x)e^{-y}).$$

Then by direct calculation F^2 can be written as

$$F^2(x, y) = (h^2(x)e^{g_1(x,y)}, \quad ye^{g_2(x,y)}).$$

The map $h(x)$ of exponential type is unimodal (cf. Example 1.3 in Chap. 1), which has been studied extensively in the discrete dynamical systems literature. The function $h(x)$ has two fixed points 0 and 1. It is strictly increasing on $[0, \frac{1}{\mu_1}]$ and strictly decreasing on $[\frac{1}{\mu_1}, +\infty)$, and

$$\lim_{x \to +\infty} h(x) = 0.$$

Thus, $h(x)$ has a global maximum value M at $x = \frac{1}{\mu_1}$, with

$$M \triangleq h(\frac{1}{\mu_1}) = \frac{1}{\mu_1} e^{\mu_1 - 1}.$$

Therefore, when $\mu_1 > 1$, there exist r_1 and r_2 with $0 < r_1 < 1 < r_2$ such that

$$h^2(r_1) = h^2(r_2) = h\left(\frac{1}{\mu_1}\right) = M. \tag{9.19}$$

See the graphics of h and h^2 in Fig. 9.1.

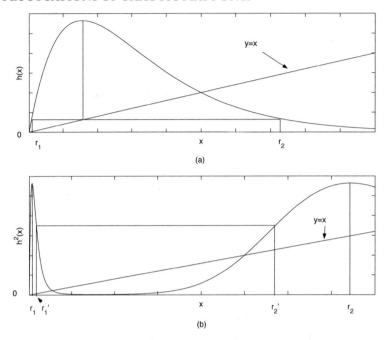

Figure 9.1: Graphics of the map $h(x)$ of the exponential type and $h^2(x)$ with $\mu_1 = 3.2$, cf. h defined in (9.18).

In Theorem 9.24 below, we show that F^2 has a horseshoe. Therefore, the map F is chaotic. $\square$

Lemma 9.22   There exists a constant μ_1^0 such that when $\mu_1 > \mu_1^0$, we have

$$h^3\left(\frac{1}{\mu_1}\right) < \frac{1}{\mu_1}. \tag{9.20}$$

Thus, if $\mu_1 > \mu_1^0$, then

$$h^2\left(\frac{1}{\mu_1}\right) < r_1.$$

Proof. A routine calculation shows that

$$h^3\left(\frac{1}{\mu_1}\right) = \frac{1}{\mu_1}\exp(3\mu_1 - 1 - e^{\mu_1-1} - e^{2\mu_1-1-e^{\mu_1-1}}).$$

We see that (9.20) holds if and only if

$$3\mu_1 - 1 - e^{\mu_1-1} - e^{2\mu_1-1-e^{\mu_1-1}} < 0.$$

A sufficient condition for the above inequality is

$$3\mu_1 - 1 - e^{\mu_1-1} < 0.$$

This is obviously true if μ_1 is large enough. □

Remark 9.23 Numerical computations show that (9.20) holds when $\mu_1 > 3.117$. □

In the following, we fix $\mu_1 > \mu_1^0$. Denote by p_2 the period-2 point of h in $(r_1, \frac{1}{\mu_1})$. From (9.19) and the unimodal properties of $h(x)$, it follows that for any $1 < r_2' < r_2$, there exists a unique r_1' with $r_1 < r_1' < p_2$ such that

$$h^2(r_1') = h^2(r_2') > r_2'. \tag{9.21}$$

For any $\varepsilon > 0$, let

$$Q_\varepsilon \equiv [r_1', r_2'] \times [0, \varepsilon], \tag{9.22}$$

and

$$M_1 = \max_{r_1 < x < r_2} \{x + h(x)\}. \tag{9.23}$$

Theorem 9.24 Let $\mu_1 > \mu_1^0$, where μ_1^0 is given in Lemma 9.22. If

$$0 < s < \frac{2\mu_2}{M_1}, \tag{9.24}$$

then for any r_2' with $1 < r_2' < r_2$, there exists an $\varepsilon > 0$, such that the map $F^2 \colon Q_\varepsilon \to \mathbb{R}^2$ has a topological horseshoe with a crossing number $m = 2$.

Proof. Let r_1' and r_2' be defined as above. Since $\mu_1 > \mu_1^0$, there exist r_3, r_4 such that

$$p_2 < r_3 < \frac{1}{\mu_1} < r_4 < 1, \tag{9.25}$$

$$h^2(r_3) = h^2(r_4) < r_1 < r_1'. \tag{9.26}$$

On the other hand, we have

$$g_1(x, 0) = 0,$$

for any $x > 0$.

It follows from (9.21) and (9.26) that there exists an $\varepsilon > 0$ such that

$$h^2(r_1')e^{g_1(r_1',y)} > r_2', \quad h^2(r_2')e^{g_1(r_2',y)} > r_2', \tag{9.27}$$
$$h^2(r_3)e^{g_1(r_3,y)} < r_1', \quad h^2(r_4)e^{g_1(r_4,y)} < r_1', \tag{9.28}$$

for any $y \in [0, \varepsilon]$.

On the other hand, we have

$$ye^{g_2(x,y)} < y, \tag{9.29}$$

for any $y > 0$ and $r_1 < x < r_2$.

Denote

$$Q_1 = \{(x, y) \mid x = r_1', \, 0 \le y \le \varepsilon\}, \qquad Q_2 = \{(x, y) \mid x = r_2', \, 0 \le y \le \varepsilon\},$$
$$D_1 = \{(x, y) \mid r_1' \le x \le r_3, \, 0 \le y \le \varepsilon\}, \quad D_2 = \{(x, y) \mid r_4 \le x \le r_2', \, 0 \le y \le \varepsilon\}.$$

If we denote

$$(\bar{x}, \bar{y}) = F^2(x, y),$$

then by (9.27)–(9.29), for any $(\bar{x}, \bar{y}) \in F^2(Q_i), \, i = 1, 2$, we have $\bar{x} > r_2'$ and $0 < \bar{y} < \varepsilon$. Likewise, by the same argument, for any $(\bar{x}, \bar{y}) = F^2(\{r_3\} \times [0, \varepsilon]), \, i = 1, 2$, we have $\bar{x} < r_1'$ and $0 < \bar{y} < \varepsilon$. Thus, if Γ is a connection, then by our discussion above we can see that $\gamma_i \equiv \Gamma \cap D_i \, (i = 1, 2)$ are two mutually disjoint preconnections, since the curve $F^2(\gamma_i)$ crosses Q_1 and $Q_2, \, i = 1, 2$. See Fig. 9.2. The proof is complete. $\quad\square$

From the definition of Q_ε, there exists a connection Γ_0 of Q_ε such that

$$\dim_{\mathcal{H}}(\Gamma_0) = 1, \qquad \mathcal{H}^1(\Gamma_0) = r_2' - r_1'.$$

Thus, under the assumption of Theorem 8.8, the system $(\mathbb{R}^2, f^2)$ has rapid fluctuations of dimension 1 by Theorem 7.3, and so does $(\mathbb{R}^2, f)$ by the Lipschitz property of f.

Exercise 9.25   Let $f \colon D \subset \mathbb{R}^N \to D$ be Lipschitz continuous with Lipschitz constant L and A be an s-set in D. Prove that for any nonnegative integer k, we have

$$\mathrm{Var}_{f(D)}^{(s)}(f^{n+k-1}) \le \mathrm{Var}_D^{(s)}(f^{n+k}) \le L^{sk}\,\mathrm{Var}_D^{(s)}(f^n), \qquad \forall n \ge 1. \quad\square$$

Exercise 9.26   Let $f \colon D \subset \mathbb{R}^N \to D$ be Lipschitz continuous with Lipschitz constant L and A be an s-set in D. Prove that

$$\limsup_{n\to\infty} \frac{1}{n} \ln \mathrm{Var}_D^{(s)}(f^n) = \lim_{n\to\infty} \frac{1}{n} \ln \mathrm{Var}_D^{(s)}(f^n). \quad\square$$

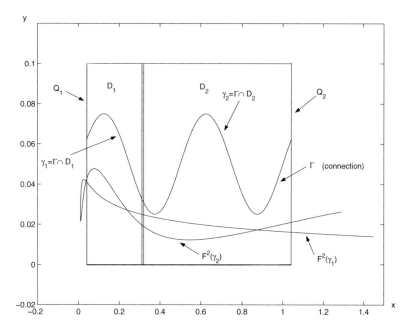

Figure 9.2: The existence of a topological horseshoe for f^2 with $\mu_1 = 3.2, \mu_2 = 1, s = 0.6$ and $\varepsilon = 0.1$.

Exercise 9.27   Let $f : [0, 1] \to \mathbb{R}$ be continuously differentiable. Show that

$$V_{[0,1]}(f) = \int_0^1 |f'(x)| dx.$$
□

Exercise 9.28   Under the assumption in Exercise 9.26, if

$$\lim_{n \to \infty} \mathrm{Var}_D^{(s)}(f^n) = \infty,$$

show that there exists $x \in D$, such that

$$\lim_{n \to \infty} \mathrm{Var}_{D \cap U_\varepsilon(x)}^{(s)}(f^n) = \infty,$$

where $U_\varepsilon(x)$ denotes a closed ε-neighborhood of x.
□

Exercise 9.29   Let $f\colon [0, 1] \to [0, 1]$ be defined as

$$f(x) = \begin{cases} 3x & 0 \le x \le \dfrac{1}{3} \\[2mm] 1 & \dfrac{1}{3} < x \le \dfrac{2}{3} \\[2mm] -3x + 3 & \dfrac{2}{3} < x \le 1. \end{cases}$$

Show that f has rapid fluctuations of dimension s for any $s \in (0, 1]$.   □

Exercise 9.30   Under the assumption in Exercise 9.25, if f has rapid fluctuations on an s-set A, show that there exists at least an $x \in A$ such that

$$\lim_{n\to\infty} \frac{1}{n} \operatorname{Var}_A^{(s)}(f^n) = \lim_{n\to\infty} \frac{1}{n} \operatorname{Var}_{A\cap U_\varepsilon(x)}^{(s)}(f^n), \qquad \forall \varepsilon > 0,$$

where $U_\varepsilon(x)$ denotes a closed ε-neighborhood of x.   □

Exercise 9.31   Let $f\colon D \subset \mathbb{R}^N \to D$ be Lipschitz continuous with Lipschitz constant L and A be an s-set in D. Prove that

$$\lim_{n\to\infty} \frac{1}{n} \ln \operatorname{Var}_A^{(s)}(f^n) = \lim_{n\to\infty} \frac{1}{n} \ln \sum_{i=1}^{n} \operatorname{Var}_A^{(s)}(f^i).$$

(Here we set $0 \ln 0 = 0$.)   □

Exercise 9.32   Prove the assertion in Example 9.13.   □

NOTES FOR CHAPTER 9

This chapter is a continuation of Chapter 2. The notion of rapid fluctuations was first introduced in Huang, Chen and Ma [37], motivated by the study of chaotic vibration of the wave equation; see Appendix B. More results are developed in [39, 41]. Sections 9.3–9.4 follow from [41]. In particular, we have chosen Example 9.5 from [41] and Example 9.6 from [39].

In the study of chaotic dynamics of maps on one or higher dimensions, it is worth emphasizing that chaos often happens on an invariant set (the "strange attractor") with a fractal dimension. Thus, fractional Hausdorff dimensions and Hausdorff measures studied in Chapter 7 have found nice applications to the characterization of chaos.

CHAPTER 10

Infinite-dimensional Systems Induced by Continuous-Time Difference Equations

10.1 I3DS

Let I be a closed interval in $\mathbb{R}$, $C([0, 1], I)$ be the space of continuous functions from $[0, 1]$ to I and f be a continuous map from I into itself. That is, (I, f) is a one-dimensional discrete dynamical system. In this section, we consider the following map:

$$F: \ C([0, 1], I) \longrightarrow C([0, 1], I), \quad F(\varphi) = f \circ \varphi \text{ for } \varphi \in C([0, 1], I). \quad (10.1)$$

This F maps from the infinite-dimensional space $C([0, 1], I)$ into itself and will be denoted as $(C([0, 1], I), F)$. It constitutes an *infinite-dimensional discrete dynamical system (I3DS)*. The system $(C([0, 1], I), F)$ is an infinite-dimensional dynamical system generated by the one-dimensional system (I, f).

One of the motivations of studying the system (10.1) is from the continuous-time difference equation

$$x(t + 1) = f(x(t)), \qquad t \in \mathbb{R}^+, \quad (10.2)$$

where f is a continuous map from I into itself. For each initial function φ from $[0, 1) \to I$, the difference equation (10.2) has a unique solution $x_\varphi \colon \ R^+ \to I$, which can be defined step by step:

$$x_\varphi(t) = (f^i \circ \varphi)(t - i) \quad \text{for} \quad t \in [i, i + 1), \quad i = 0, 1, \dots, \quad (10.3)$$

where f^i represents the i-th iterates of f. If only continuous solutions of (10.2) are admitted, we need *consistency conditions*

$$\varphi \in C([0, 1), I) \quad \text{and} \quad \varphi(1^-) = f(\varphi(0)). \quad (10.4)$$

If we define $\varphi(1) = \varphi(1^-)$, then $\varphi \in C([0, 1], I)$. We denote by $C^*([0, 1], I)$ the set of all such initial functions.

For $\varphi \in C^*([0, 1], I)$, there is a one-to-one correspondence between the solutions of the difference equation (10.2) and the orbits of the infinite-dimensional system (10.1).

$$x_\varphi(t + n) = F^n(\varphi)(t) \quad \text{for} \quad n \in \mathbb{Z}^+ \text{ and } t \in [0, 1], \quad (10.5)$$

where and throughout $\mathbb{Z}^+$ denotes the set of all positive integers.

The difference equation (10.2) as well as the I3DS (10.1) arise from concrete problems in applications. In particular, they are also related to nonlinear boundary value problems related to wave propagation, see [12]–[15], [63]. We will give illustrative examples in Appendix B.

10.2   RATES OF GROWTH OF TOTAL VARIATIONS OF ITERATES

We will characterize the asymptotic behavior and complexity of the solutions by means of the growth rates of the total variations. This approach is quite intuitive and natural for such a purpose. Let $BV([0, 1], I)$ be the space of all functions from $[0, 1]$ to I with bounded total variations and $V_{[0,1]}(\varphi)$ be the same as in (2.1). To study the asymptotic behavior of the I3DS (10.1), we will be concerned with the growth rates of $V_{[0,1]}(F^n(\varphi))$ as $n \to \infty$. The following are three distinct cases as n tends to ∞:

(1) $V_{[0,1]}(F^n(\varphi))$ remains bounded;

(2) $V_{[0,1]}(F^n(\varphi))$ grows unbounded, but may or may not be exponentially with respect to n;

(3) $V_{[0,1]}(F^n(\varphi))$ grows exponentially with respect to n.

Those properties are decided completely by the one-dimensional dynamical system (I, f). Thus, we must consider the relationship between the growth rates of the total variations of f^n as $n \to \infty$ and the complexity of (I, f).

Throughout the following, we assume that $f \in C(I, I)$ and f is *piecewise monotone* with finitely many extremal points. Let $PM(I, I)$ denote the set of all such maps.

Definition 10.1   Let $f \in PM(I, I)$ and $x \in I$.

(1) x is called a point of bounded variation of f if there exists a neighborhood J of x in I such that $V_J(f^n)$ remains bounded for all $n \in \mathbb{Z}^+$;

(2) x is called a point of unbounded variation of f if, for any neighborhood J of x, $V_J(f^n)$ grows unbounded as $n \to \infty$;

(3) x is called a chaotic or rapid fluctuation point of f if

$$\gamma(x, f) \equiv \lim_{\varepsilon \to 0} \gamma([x - \varepsilon, x + \varepsilon] \cap I, f) > 0,$$

where

$$\gamma(J, f) = \limsup_{n \to \infty} \frac{1}{n} \ln V_J(f^n), \tag{10.6}$$

for any subinterval J in I.

Denote by $B(f)$, $U(f)$ and $E(f)$ the sets of all, respectively, points of bounded variation, points of unbounded variation and chaotic points of f. □

It is easy to see from Definition 10.1 that $B(f)$ is open in I. Thus, $U(f)$ is closed in I and

$$B(f) \cap U(f) = \emptyset, \qquad B(f) \cup U(f) = I, \qquad E(f) \subset U(f).$$

For the orbits of the I3DS (10.1), we have the following classification.

Theorem 10.2 For an initial function $\varphi \in C([0, 1], I)$, let $R(\varphi)$ denote the range of φ and $\dot{R}(\varphi)$ the interior of $R(\varphi)$.

(1) If $B(f) \neq \emptyset$, $R(\varphi) \subset B(f)$ and φ is piecewise monotone with finitely many extremal points, then $V_{[0,1]}(F^n(\varphi))$ remains bounded for all $n \in \mathbb{Z}^+$.

(2) If $U(f) \neq \emptyset$ and $U(f) \cap \dot{R}(\varphi) \neq \emptyset$, then $V_{[0,1]}(F^n(\varphi))$ grows unbounded as $n \to \infty$.

(3) If $E(f) \neq \emptyset$ and $E(f) \cap \dot{R}(\varphi) \neq \emptyset$, then $V_{[0,1]}(F^n(\varphi))$ grows exponentially with respect to n as $n \to \infty$.

Proof. For case (1), since φ is continuous and $B(f)$ is open, there exists an interval $J \subset B(f)$ such that $R(\varphi) \subset J$ by assumptions. Let $\ell(\varphi)$ denote the number of maximal closed subintergals of $[0, 1]$ on each of which φ is monotonic. (Call this number the *lap* of φ.) Then we have

$$V_{[0,1]}(F^n(\varphi)) \leq \ell(\varphi) V_J(f^n).$$

Thus, $V_{[0,1]}(F^n(\varphi))$ remains bounded for all $n \in \mathbb{Z}^+$.

In the case of (2), it follows from the assumptions that there exist a point $x \in U(f)$ and a neighborhood J of x in I such that

$$x \in J \subset R(\varphi).$$

Thus,

$$V_{[0,1]}(F^n(\varphi)) \geq V_J(f^n), \qquad \forall n \in \mathbb{Z}^+.$$

This implies (2).

The proof for case (3) is the same as the one for case (2). □

In the remaining three sections of this chapter, we shall consider the properties of $B(f)$, $U(f)$ and $E(f)$, respectively.

10.3   PROPERTIES OF THE SET $B(f)$

We first give the following.

Theorem 10.3   Let $f \in PM(I, I)$. Then $V_I(f^n)$ remains bounded if and only if $B(f) = I$.

Proof. Necessity follows directly from the definition.

Now we prove the sufficiency. Since $B(f) = I$, it follows that for each $x \in I$ there exists an open neighborhood U_x of x such that

$$V_{U_x}(f^n) \le C_x$$

for some $C_x > 0$ which is independent of n.

Since $\{U_x, x \in I\}$ is an open cover of the compact interval I, there is a finite subcover such that

$$I \subset \bigcup_{i=1}^{n} U_{x_i}.$$

So

$$V_I(f^n) \le \sum_{i=1}^{n} V_{U_{x_i}}(f^n) \le C$$

where

$$C = \max_{1 \le i \le n} \{C_{x_i}\}. \tag{10.7}$$

□

Definition 10.4   ([63]) A continuous map $g: I = [a, b] \to I$ is called an L-map if there exists $[x_1, x_2] \subseteq I$, $x_1 \le x_2$ such that

$$g(x) = x, \ \text{for } x \in [x_1, x_2], \tag{10.8}$$
$$x < g(x) \le x_2, \ \text{for } x < x_1, \ \text{and } g(x) < x, \ \text{for } x > x_2. \tag{10.9}$$

□

In [63], it is shown that every orbit of the I3DS (10.1) is compact and hence possesses a nonempty compact ω-limit set in $C([0, 1], I)$ if and only if either f or f^2 is an L-map.

Theorem 10.5   Let $f \in C(I, I)$ be piecewise monotone with finitely many critical points. If either f or f^2 is an L-map, then $B(f) = I$.

Proof. It suffices to show that $V_I(f^n)$ remains bounded for all $n \in \mathbb{Z}^+$. Without loss of generality, we assume that $a < x_1$, where x_1 (and x_2) are defined according to Definition 10.4. Here we merely need to prove that $V_{[a,x_1]}(f^n)$ remains bounded for all $n \in \mathbb{Z}^+$ since $V_{[x_1,x_2]}(f^n)$ remains constant for all $n \in \mathbb{Z}^+$, and the boundedness of $V_{[x_2,b]}(f^n)$ can be proved similarly.

If f is an L-map, then f has only fixed points rather than periodic points. Since f has finitely many critical points and is piecewise monotone, there exists a $\delta_0 > 0$ such that $[x_1 - \delta_0, x_1] \subset I$ and f is monotonic on $[x_1 - \delta_0, x_1]$. By (10.8) and (10.9), f^n is also monotonic on $[x_1 - \delta_0, x_1]$ for every $n \in \mathbb{Z}^+$. On the other hand, $\lim_{n\to\infty} f^n(x)$ converges to a fixed point of f uniformly in $x \in I$ since f is an L-map. Thus, there exists a positive integer N such that when $n \geq N$,

$$f^n(x) > x_1 - \delta, \qquad \forall x \in [a, x_1], \tag{10.10}$$

due to the fact that f has no fixed points in $[a, x_1)$.

Since $f \in PM(I, I)$, so is f^n for any $n \in \mathbb{Z}^+$. Let $\ell(f)$ denote the number of different laps on which f is strictly monotone and let

$$l_N = \max\{\ell(f^N), \ell(f^{N+1}), \ldots, \ell(f^{2N-1})\}.$$

Consider the map f^N. Let J_1 and J_2 be, respectively, the subintervals in $[a, x_1]$ such that

$$f^N(x) \leq x_1, \quad \text{for} \quad x \in J_1,$$
$$(x_2 \geq) f^N(x) \geq x_1, \quad \text{for} \quad x \in J_2.$$

Then $J_1 \cup J_2 = [a, x_1]$ and for any $n \in \mathbb{Z}^+$, $f^{Nn}(x) = f^N(x)$ for all $x \in J_2$, and the lap number of f^{Nn} on J_1 does not increase with n increasing. Thus,

$$
\begin{aligned}
V_{[a,x_1]}(f^{Nn}) &= V_{J_1}(f^{Nn}) + V_{J_2}(f^{Nn}) \\
&\leq \ell(f^N)|f(x_1 - \delta) - f(x_1)| + V_{J_2}(f^N) \\
&\leq \ell_N|f(x_1 - \delta) - x_1| + V_{[a,x_1]}(f^N).
\end{aligned}
\tag{10.11}
$$

Similarly, we have

$$
\begin{aligned}
V_{[a,x_1]}(f^{(N+i)n}) &\leq \ell(f^{N+i})|f(x_1 - \delta) - x_1| + V_{[a,x_1]}(f^{N+i}) \\
&\leq \ell_N|f(x_1 - \delta) - x_1| + V_{[a,x_1]}(f^N),
\end{aligned}
\tag{10.12}
$$

for $i = 1, 2, \ldots, N - 1$. Inequalities (10.11) and (10.12) imply that the total variation $V_{[a,x_1]}(f^n)$ of f^n on $[a, x_1]$ remains bounded for all $n \in \mathbb{Z}^+$.

If f^2 is an L-map but f itself is not, then f has periodic points of period less than or equal to 2 only. Utilizing the same argument, we can derive that $V_I(f^{2n})$ remains bounded for all $n \in \mathbb{Z}^+$. Note that

$$V_I(f^{2n+1}) \leq \ell(f)V_I(f^{2n}),$$

hence, we have that $V_I(f^n)$ remains bounded for all $n \in \mathbb{Z}^+$. This completes the proof of the theorem. $\square$

Remark 10.6   The converse of Theorem 10.5 need not be true. That is, there are maps in $PM(I, I)$ which are not L-maps but their total variations remain bounded for all $n \in \mathbb{Z}^+$. For instance, $f(x) = \sqrt{x}$ on $[0, 1]$. It is easy to see that $V_{[0,1]}(f^n) = 1$ for all $n \in \mathbb{Z}^+$, but neither f nor f^2 is an L-map. Thus, not every orbit of the I3DS (10.1) defined through such an f is compact. We will see from Theorem 10.20 below that a necessary condition for $B(f) = I$ is that f has no periodic orbit with period great than 2.   □

Theorem 10.7   Let $f \in PM(I, I)$. Then we have

(1) If $x \in \mathrm{Fix}(f)$, and x is a local stable point of f, then $x \in B(f)$;

(2) If f has at most a periodic point in $\mathrm{int}(I)$, then $B(f) = I$.   □

Exercise 10.8   Prove Theorem 10.7.   □

10.4   PROPERTIES OF THE SET $U(f)$

We have the following.

Theorem 10.9   Let $f \in PM(I, I)$. Then $\lim_{n\to\infty} V_I(f^n) = \infty$ if and only if $U(f) \neq \emptyset$.

Proof.   The sufficiency immediately follows by definition. To prove the necessity, we assume that

$$\lim_{n\to\infty} V_I(f^n) = \infty.$$

Let $I = [a, b]$. Consider the two bisected subintervals $[a, (a + b/2)]$ and $[(a + b)/2, b]$. Then the total variations of f^n grow unbounded at least on one of the two intervals as $n \to \infty$. We denote by $I_1 = [a_1, b_1]$ such an interval. Repeating the above bisecting processing, we obtain a sequence of intervals $I_k = [a_k, b_k]$ with

(i) $[a, b] \supset [a_1, b_1] \supset \cdots \supset [a_k, b_k] \supset \cdots$,

(ii) $b_k - a_k = \frac{b-a}{2^k}$,

(iii) $V_{I_k}(f^n)$ grows unbounded as $n \to \infty$ for each $k \in \mathbb{Z}^+$.

It follows from (i) and (ii) that there exists a unique $c \in I_k$ for all k such that

$$\lim_{k\to\infty} a_k = \lim_{k\to\infty} b_k = c.$$

For any neighborhood J of c, there exists a $k \in \mathbb{Z}^+$ such that $J \supset I_k$. Following (iii), we have that $c \in U(f)$.   □

The following is the main theorem of this section.

Theorem 10.10   Let $f \in PM(I, I)$. Then $U(f) = I$ if and only if f has sensitive dependence on initial data on I.   □

To prove the sufficiency of Theorem 10.10, we need the following lemma.

Lemma 10.11   Assume that $f \in PM(I, I)$ and f has sensitive dependence on initial data on I. Let $J = [c, d]$ be an arbitrary subinterval of I, with $|J| \geq \delta$, where δ is the sensitive constant of f. Then there exists an $A: 0 < A \leq \delta/2$, independent of J, such that

$$|f^n(J)| \geq A, \qquad \forall n \in \mathbb{Z}^+. \tag{10.13}$$

Proof. Let $N = [\frac{2(b-a)}{\delta}] + 1$, where $[r]$ denotes the usual integral part of the real number r. Divide I into N equal-length subintervals I_i, $i = 1, \ldots, N$, with $I_i = [x_{i-1}, x_i]$, $x_0 = a$ and $x_N = b$, $|I_i| = (b-a)/N$. Then $|I_i| \leq \delta/2$ holds for each i.

Let $x \in \text{int}(I_i) = (x_{i-1}, x_i)$, the interior of I_i. Then from the sensitive dependence on initial data of f (cf. Definition 6.27), it follows that there is a $y \in \text{int}(I_i)$ and an $N_i \in \mathbb{Z}^+$ such that

$$|f^{N_i}(x) - f^{N_i}(y)| \geq \delta.$$

This implies

$$|f^{N_i}(I_i)| \geq \delta.$$

For $i = 1, \ldots, N$, let

$$a_i = \min\{|f^j(I_i)| \mid j = 0, 1, \ldots, N_i\}, \tag{10.14}$$

and

$$A = \min\{a_i \mid i = 1, \ldots, N\}; \qquad A > 0. \tag{10.15}$$

Then $A \leq \delta/2$ because $|I_i| \leq \delta/2$ for each i. Now let $J = [c, d]$ satisfy $|J| \geq \delta$. Then there exists at least a subinterval I_{j_0}, $1 \leq j_0 \leq N$, such that $J \supset I_{j_0}$. Thus,

$$f^k(J) \supset f^k(I_{j_0}), \qquad \text{for } k = 0, 1, \ldots. \tag{10.16}$$

We are ready to establish (10.13). We divide the discussion into the following cases:

(i) $0 \leq n \leq N_{j_0}$. In this case, it is obvious that

$$|f^n(J)| \geq |f^n(I_{j_0})| \geq a_{j_0} \geq A,$$

by (10.14)–(10.16). So (10.13) holds.

(ii) $n > N_{j_0}$. If $0 < n - N_{j_0} \leq \min\{N_i \mid i = 1, \ldots, N\}$, then from the facts that $f^{N_{j_0}}(J) \supset f^{N_{j_0}}(I_{j_0})$ and $f^{N_{j_0}}(I_{j_0})$ having length at least δ and then further containing at least one subinterval I_{j_1}, we obtain

$$f^{n-N_{j_0}}(f^{N_{j_0}}(J)) \supset f^{n-N_{j_0}}(f^{N_{j_0}}(I_{j_0})) \supset f^{n-N_{j_0}}(I_{j_1}). \qquad (10.17)$$

But $n - N_{j_0} \leq \min\{N_i \mid i = 1, \ldots, N\}$. By (10.14), (10.15) and (10.17), we have

$$|f^n(J)| = |f^{n-N_{j_0}}(f^{N_{j_0}}(J))| \geq |f^{n-N_{j_0}}(I_{j_1})| \geq A.$$

The above restriction that $0 < n - N_{j_0} \leq \min\{N_i \mid i = 1, 2, \ldots, N\}$ can actually be relaxed to $0 < n - N_{j_0} \leq N_{j_1}$, if I_{j_1} is the subinterval satisfying (10.17), by (10.14) and (10.15).

One can then extend the above arguments inductively to any $n = N_{j_0} + N_{j_1} + \cdots + N_{j_k} + R_k$, where $R_k \in \mathbb{Z}^+ \cup \{0\}, 0 \leq R_k \leq N_{j_{k+1}}$, and where

$$f^n(J) = f^{R_k}(f^{N_{j_0}+\cdots+N_{j_k}}(J)) \supseteq f^{R_k}(f^{N_{j_k}}(I_{j_k})) \supseteq f^{R_k}(I_{j_{k+1}}) \qquad (10.18)$$

is satisfied for a sequence of intervals $I_{j_0}, I_{j_1}, \ldots, I_{j_k}$ and $I_{j_{k+1}}$. From (10.14), (10.15) and (10.18), we have proved (10.13). $\qquad\qquad\square$

Equipped with Lemma 10.11, we can now proceed to give the following.

Proof of the sufficiency of Theorem 10.10. Let $J = [c, d]$ be any subinterval of I and let $M > 0$ be sufficiently large. We want to prove the following statement:

"there exists an $N(M) \in \mathbb{Z}^+$ such that $V_J(f^n) \geq M$ for all $n \geq N(M)$". $\qquad (10.19)$

First, divide J into N subintervals, with $N = \left[\frac{M}{A}\right] + 1$, where A satisfies Lemma 10.11. Thus, $J = J_1 \cup J_2 \cup \cdots \cup J_N$, with $J_k = [x_{k-1}, x_k]; x_k = c + k\left(\frac{d-c}{N}\right), k = 1, 2, \ldots, N$. By the sensitive dependence of f on I, for any $k = 1, 2, \ldots, N$, there exists an N_k such that

$$|f^{N_k}(J_k)| \geq \delta. \qquad (10.20)$$

Since $f^{N_k}(J_k)$ is a connected interval, by (10.20) we can apply Lemma 10.11 and obtain

$$|f^n(J_k)| = |f^{n-N_k}(f^{N_k}(J_k))| \geq A, \text{ if } n \geq N_k, \text{ for } k = 1, 2, \ldots, N.$$

Now take $N(M) = \max\{N_1, \ldots, N_N\}$. Then for $n \geq N(M)$,

$$V_J(f^n) = \sum_{k=1}^{N} V_{J_k}(f^n) \geq \sum_{k=1}^{N} |f^n(J_k)| \geq \sum_{k=1}^{N} A = NA > M.$$

The proof of (10.9) and, therefore, the sufficiency of Theorem 10.10 are now established. $\qquad\square$

For the necessity of Theorem 10.10, a long sequence of proposition and lemmas is needed in order to address the case $U(f) = I$, i.e., every point $x \in I$ is a point of unbounded variation.

Proposition 10.12 Assume $f \in PM(I, I)$ and $U(f) = I$. Then

(i) $f(x) \not\equiv c$ on any subinterval J of I, for any constant c;

(ii) $f(x) \not\equiv x$ on any subinterval J of I;

(iii) let J be a subinterval of I whereupon f is monotonic. Then there exists at most one point $\bar{x} \in J$ such that $f(\bar{x}) = \bar{x}$. Consequently, f has at most finitely many fixed points on I;

(iv) let J be a subinterval of I and $x_0 \in J$ satisfies $f(x_0) = x_0$. If f is increasing on J, then $f(x) > x$ for all $x > x_0, x \in J$, and $f(x) < x$ for all $x < x_0, x \in J$. This property also holds if J is an interval with x_0 either as its left or right endpoint:

(v) for any positive integer n, $f^n \in PM(I, I)$ and $U(f^n) = I$.

Proof. Part (i) is obvious. Consider part (ii). If $f(x) \equiv x$ on J, then

$$V_J(f^n) = |J| \quad \text{for every} \quad n .$$

This violates $U(f) = I$.

Now consider (iii). Let us first assume that f is monotonically decreasing on J. Define $g(x) = f(x) - x$. Then g is also decreasing on J. If there were two points $\bar{x}_1$ and $\bar{x}_2; \bar{x}_1, \bar{x}_2 \in J$, $\bar{x}_1 \neq \bar{x}_2$, such that $f(\bar{x}_i) = \bar{x}_i, i = 1, 2$, then $g(\bar{x}_1) = g(\bar{x}_2)$ and therefore $g(x) \equiv 0$ on a subinterval of J, implying $f(x) \equiv x$ on J, contradicting part (ii).

If f is monotonically increasing on J and there exist two fixed points $\bar{x}_1, \bar{x}_2 \in J, \bar{x}_1 < \bar{x}_2$, then f is monotonically increasing on $J_0 \equiv [\bar{x}_1, \bar{x}_2]$, and f^n is also increasing on J_0, such that $f^n(J_0) = J_0$ for every $n \in \mathbb{Z}^+$. Thus, $V_{J_0}(f^n) = |J_0| \nrightarrow \infty$ as $n \to \infty$, contradicting $U(f) = I$. Therefore, we have established (iii).

Further, consider (iv). If there exists an $x_1 \in J$ such that $x_1 > x_0$ and $f(x_1) \leq x_1$, then $f(x_1) < x_1$ because $f(x_1) = x_1$ is ruled out by (iii). Consider $J_1 \equiv [x_0, x_1]$, if $x_0 < x_1$. Then f is increasing on J_1, so are f^n for any $v \in \mathbb{Z}^+$, such that $f^n(J_1) \subseteq J_1$. Hence, $V_{J_1}(f^n) \leq |J_1|$, violating $U(f) = I$. The case that $x_1 < x_0$ and $f(x_1) \geq x_1$ similarly also leads to a contradiction.

Finally, part (v) is obvious. Its proof is omitted. □

Remark 10.13 We see that Proposition 10.12 (iv) is actually a *hyperbolicity* result, i.e., if x_0 is a fixed point of f and $x_0 \in U(f)$, and if f is increasing and differentiable at x_0, then $|f'(x_0)| > 1$.□

Lemma 10.14   Assume $f \in PM(I, I)$ and $U(f) = I$. Let $\bar{x}_0$ be a fixed point of f on I and U be a small open neighborhood of $\bar{x}_0$ in I. Then there exists a $\delta_0 > 0$ such that for any $x \in U \backslash \{\bar{x}_0\}$, there exists an $N_x \in \mathbb{Z}^+$, N_x depending on x, such that

$$|f^{N_x}(x) - \bar{x}_0| > \delta_0. \tag{10.21}$$

Proof. Since $f \in PM(I, I)$, we have two possibilities: (i) f is monotonic on $U = [\bar{x}_0 - \delta, \bar{x}_0 + \delta]$ for some sufficiently small $\delta > 0$; (ii) $\bar{x}_0$ is an extremal point of f.

First, consider case (i) when f is increasing on U. Since $U(f) = I$, Proposition 10.12 part (iv) gives

$$f(\bar{x}_0 - \delta) < \bar{x}_0 - \delta, \quad f(\bar{x}_0 + \delta) > \bar{x}_0 + \delta .$$

Thus, we can find $x_1 \in (\bar{x}_0 - \delta, \bar{x}_0)$, $x_2 \in (\bar{x}_0, \bar{x}_0 + \delta)$, such that

$$f(x_1) = \bar{x}_0 - \delta, \quad f(x_2) = \bar{x}_0 + \delta. \tag{10.22}$$

Define $\delta_0 = \min\{\bar{x}_0 - x_1, x_2 - \bar{x}_0\}$. We now show that (10.21) is true.

Assume the contrary that (10.21) fails for some $\hat{x} \in (\bar{x}_0 - \delta, \bar{x}_0) \cup (\bar{x}_0, \bar{x}_0 + \delta)$. Then

$$|f^n(\hat{x}) - \bar{x}_0| \leq \delta_0, \quad \text{for all} \quad n \in \mathbb{Z}^+. \tag{10.23}$$

We consider the case $\hat{x} > \bar{x}_0$. (The case $\hat{x} < \bar{x}_0$ can be similarly treated and is therefore omitted.) From (10.23), $f(\bar{x}_0) = \bar{x}_0$, and the fact that f^n is increasing on $[\bar{x}_0, \bar{x}_0 + \delta_0]$, we have $f^n([\bar{x}_0, \hat{x}]) \subseteq [\bar{x}_0, \bar{x}_0 + \delta_0]$ and thus $V_{[\bar{x}_0, \hat{x}]}(f^n) \leq \delta_0$ for any $n \in \mathbb{Z}^+$, violating $U(f) = I$.

Next, consider case (i) when f is decreasing on U. By the continuity of f and Proposition 10.12 part (i), we have $f(\bar{x}_0 - \delta) > \bar{x}_0$ and $f(\bar{x}_0 + \delta) < \bar{x}_0$. Thus, we can find $x_1 \in (\bar{x}_0 - \delta, \bar{x}_0)$ and $x_2 \in (\bar{x}_0, \bar{x}_0 + \delta)$ such that

$$f(x_1) = \bar{x}_0 + \delta, \quad f(x_2) = \bar{x}_0 - \delta .$$

Let $\delta_0 = \min\{x_2 - \bar{x}_0, \bar{x}_0 - x_1\}$. If (10.21) were not true for this δ_0, then there is an $\hat{x} \in (\bar{x}_0 - \delta_0, \bar{x}_0) \cup (\bar{x}_0, \bar{x}_0 + \delta_0)$ such that

$$|f^n(\hat{x}) - \bar{x}_0| < \delta_0, \quad \text{for all} \quad n \in \mathbb{Z}^+. \tag{10.24}$$

Again, we may assume that $\hat{x} > \bar{x}_0$. (The case $\hat{x} < \bar{x}_0$ can be treated similarly.) Since f^2 is increasing on $[\bar{x}_0, \hat{x}]$ and by (10.24) and $f(\bar{x}_0) = \bar{x}_0$, we have $f^{2n}([\bar{x}_0, \hat{x}]) \subseteq [\bar{x}_0, \bar{x}_0 + \delta_0]$ for all $n \in \mathbb{Z}^+$. Therefore,

$$V_{[\bar{x}_0, \hat{x}]}(f^{2n}) \leq \delta_0, \quad \text{for any} \quad n \in \mathbb{Z}^+ ,$$

contradicting $U(f) = I$. So case (i) implies (10.21).

We proceed to treat case (ii), i.e., $\bar{x}_0$, as a fixed point of f, is also an extremal point of f. Note that it is also possible that $\bar{x}_0 = a$ or $\bar{x}_0 = b$, i.e., $\bar{x}_0$ is a boundary extremal point. Let us divide the discussion into the following four subcases: (1) $\bar{x}_0 = a$; (2) $\bar{x}_0 = b$; (3) $\bar{x}_0 \in (a, b)$ is a relative maximum; and (4) $\bar{x}_0 \in (a, b)$ is a relative minimum.

Subcase (1) implies that $\bar{x}_0 = a$, as a fixed point, must be a local minimum. Let

$$\tilde{x}_1 = \min\{\tilde{x} | \tilde{x} \text{ is an extremal point}, \tilde{x} > \bar{x}_0\} .$$

Then by Proposition 10.12 part (iv), we have $f(\tilde{x}_1) > \tilde{x}_1$. Then there exists an $\hat{x}_1 \in (\bar{x}_0, \tilde{x}_1)$ such that $f(\hat{x}_1) = \tilde{x}_1$. Define $\delta_0 = \hat{x}_1 - \bar{x}_0$. Then since f is increasing on $[\bar{x}_0, \tilde{x}_1]$, the case can be treated just as in case (i) earlier.

Subcase (2) is a mirror image of subcase (1) and can be treated in the same way. So let us treat subcase (3). Let

$$\tilde{x}_1 = \max\{\tilde{x} | \tilde{x} \text{ is an extremal point}, \tilde{x} < \bar{x}_0\},$$
$$\tilde{x}_2 = \min\{\tilde{x} | \tilde{x} \text{ is an extremal point}, \tilde{x} > \bar{x}_0\}.$$

Then f is increasing on $[\tilde{x}_1, \bar{x}_0]$ and decreasing on $[\bar{x}_0, \tilde{x}_2]$. By Proposition 10.12 part (iv), we have $f(\tilde{x}_1) < \tilde{x}_1$. Therefore, there exists an $\hat{x}_1 \in (\tilde{x}_1, \bar{x}_0)$ such that $f(\hat{x}_1) = \tilde{x}_1$. If $f(\tilde{x}_2) < \tilde{x}_1$, then there is an $\hat{x}_2 = (\bar{x}_0, \tilde{x}_2)$ such that $f(\hat{x}_2) = \tilde{x}_1$. In this case, we set $\delta_0 = \min\{\bar{x}_0 - \hat{x}_1, \hat{x}_2 - \bar{x}_0\}$. If $f(\tilde{x}_2) \geq \tilde{x}_1$ then we set $\delta_0 = \bar{x}_0 - \hat{x}_1$. The remaining arguments go the same way as in (i) earlier.

Subcase (4) can be treated in the same way as Subcase (3). $\square$

Recall $\omega(x, f)$, the ω-limit set of a point x in I under f; cf. Definition 6.32.

Lemma 10.15   Let $f \in C(I, I)$ and $\hat{x} \in I$. If $\omega(\hat{x}, f) = \{x_0, \ldots, x_k\}$, then $f^k(x_i) = x_i$, for $i = 0, 1, \ldots, k$. That is, $\omega(\hat{x}, f)$ is a periodic orbit. $\square$

Exercise 10.16   Prove Lemma 10.15. $\square$

Lemma 10.17   Assume that $f \in PM(I, I)$ and $U(f) = I$. Let J be any subinterval of I. Then there exists an infinite sequence $\{n_j \in \mathbb{Z} \mid j = 1, 2, \ldots\}$, $n_j \to \infty$, such that $f^{n_j}(J)$ contains at least an extremal point of f for all n_j.

Proof. If f is not monotonic on J, take $n_1 = 0$. Then $f^{n_1}(J) = J$ contains an extremal point of f.

If f is monotonic on J, then because f cannot be constant on J, f must be either strictly increasing or strictly decreasing on J. Assume first that f is strictly increasing. Then there exists some $m_1 \geq 2$ such that f^{m_1} is not monotonic on J because otherwise

$$V_J(f^n) \leq b - a \quad \text{for all} \quad n = 1, 2, \ldots,$$

a contradiction. This implies that f is not monotonic on $f^{m_1-1}(J)$ and, therefore, $f^{m_1-1}(J)$ has an extremal point of f. We then choose $n_1 = m_1 - 1 \geq 1$ in this case. (If instead f is strictly decreasing on J, then the proof is similar.)

Since $U(f) = I$, $f^{n_1}(J)$ does not collapse to a single point by Proposition 10.12. Choose a subinterval J_1 of $f^{n_1}(J)$ where f is monotonic on J_1. Using the above arguments again, we have some $m_2 \geq 2$ such that f^{m_2} is not monotonic on J_1. Therefore, $f^{m_2-1}(J_1)$ contains an extremal point of f. But $J_1 \subseteq f^{n_1}(J)$, and so $f^{m_2-1}(J_1) \subseteq f^{n_1+m_2-1}(J)$ contains an extremal point of f. Define $n_2 = n_1 + m_2 - 1$.

This process can be continued indefinitely. The proof is complete. $\square$

Lemma 10.18   Assume $f \in PM(I, I)$ and $U(f) = I$. Let $\tilde{x}_0$ be an extremal point of f. Then there is a $\delta > 0$ such that for any (relatively) open neighborhood U of $\tilde{x}_0$, there is an $\hat{x} \in U$ and an $\widehat{N} \in \mathbb{Z}^+$ such that $|f^{\widehat{N}}(\hat{x}) - f^{\widehat{N}}(\tilde{x}_0)| \geq \delta$.

Proof. Let $E = \{\tilde{x}_0, \tilde{x}_1, \ldots, \tilde{x}_k\}$ be the set of all extremal points of f. We may note that by Proposition 10.12 part (iv) and $f \in PM(I, I)$ that we have $a, b \in E$. Consider the orbit of $\tilde{x}_0$: $\mathcal{O}(\tilde{x}_0) = \{f^n(\tilde{x}_0) \mid n = 1, 2, \ldots\}$. There are two possibilities.
Case 1:   There are n_1, n_2: $n_1 > n_2 \geq 0$, such that $f^{n_1}(\tilde{x}_0) = f^{n_2}(\tilde{x}_0)$;
Case 2:   For any $n_1, n_2 \in \mathbb{Z}^+$, $f^{n_1}(\tilde{x}_0) \neq f^{n_2}(\tilde{x}_0)$ if $n_1 \neq n_2$.

Consider Case 1 first. Let $y_0 = f^{n_2}(\tilde{x}_0)$. For any interval W, $f^{n_2}(W)$ is also an interval because f^{n_2} is continuous. This interval $f^{n_2}(U)$ can never degenerate into a point by Proposition 10.12 part (i). Set $F(x) = f^{n_1-n_2}(x)$. Then $F \in PM(I, I)$ and $U(F) = I$. Pick $y_1 \in f^{n_2}(W)$ but $y_1 \neq y_0$. Let $\hat{x}_1 \in W$ satisfy $f^{n_2}(\hat{x}_1) = y_1$. Then because y_0 is a fixed point of F, by Lemma 10.14 there are a $\delta > 0$ (independent of y_1) and an $N \in \mathbb{Z}^+$ (dependent on y_1) such that

$$|F^N(y_1) - F^N(y_0)| = |F^N(y_1) - y_0| \geq \delta,$$

or

$$|f^{N(n_1-n_2)}(y_1) - f^{N(n_1-n_2)}(y_0)| \geq \delta,$$
$$|f^{N(n_1-n_2)+n_2}(\hat{x}_1) - f^{N(n_1-n_2)+n_2}(\tilde{x}_0)| \geq \delta.$$

Therefore, Lemma 10.18 holds for Case 1.

Next, consider Case 2. For the ω-limit set $\omega(\tilde{x}, f)$, there are two subcases:
Case (2.a): $\omega(\tilde{x}, f) \not\subseteq E$;
Case (2.b): $\omega(\tilde{x}, f) \subseteq E$.

Consider Case (2.a). Let $y_0 \in \omega(\tilde{x}, f)$ but $y_0 \notin E$, and let $\delta_0 = \frac{1}{2}d(y_0, E)$. By Lemma 10.17, for W there is an $N_1 \in \mathbb{Z}^+$ and a sequence $\{n_j\}$ such that $f^{n_j}(W)$ contains at least an extremal point of f for all $n_j \geq N_1$. Since $y_0 \in \omega(\tilde{x}, f)$, there is an $n_k > N_1$ such that $|f^{n_k}(\tilde{x}_0) - y_0| < \delta_0/3$. Let $\tilde{x}_{\tilde{j}} \in E$ be such that $\tilde{x}_{\tilde{j}} \in f^{n_k}(W)$, and let $\hat{x} \in W$ be such that $f^{n_k}(\hat{x}) = \tilde{x}_{\tilde{j}}$. Then

$$
\begin{aligned}
|f^{n_k}(\tilde{x}_0) - f^{n_k}(\hat{x})| &\geq |y_0 - f^{n_k}(\hat{x})| - |y_0 - f^{n_k}(\tilde{x}_0)| \\
&\geq d(y_0, E) - |y_0 - f^{n_k}(\tilde{x}_0)| \\
&\geq \frac{1}{2}d(y_0, E) - \delta_0/3 \\
&= \delta_0/2 - \delta_0/3 = \delta_0/6.
\end{aligned}
$$

Set $\delta = \delta_0/6$ and $\widehat{N} = n_k$. Then we have

$$
|f^{\widehat{N}}(\tilde{x}_0) - f^{\widehat{N}}(\hat{x})| \geq \delta ,
$$

so Lemma 10.18 holds true.

Now consider Case (2.b). We divide this into two further subcases:

Case (2.b.i) For all $n \in \mathbb{Z}^+$, $f^n(W) \cap \omega(\tilde{x}, f) = \emptyset$;

Case (2.b.ii) There is an $n_0 \in \mathbb{Z}^+$ such that $f^{n_0}(W) \cap \omega(\tilde{x}, f) \neq \emptyset$.

Consider Case (2.b.i). Since E is finite and by Lemma 10.17, there is an $\tilde{x}_{\tilde{i}} \in E$ and a subsequence $\{n_i \in \mathbb{Z}^+ \mid i = 1, 2, \ldots\}$ such that $f^{n_i}(W)$ always contains $\tilde{x}_{\tilde{i}}$. Since $f^n(W) \cap \omega(\tilde{x}, f) = \emptyset$, $\tilde{x}_{\tilde{i}} \notin \omega(\tilde{x}, f)$. Let $\delta = \frac{1}{2}d(\tilde{x}_{\tilde{i}}, \omega(\tilde{x}, f)) > 0$. Since $\lim_{n \to \infty} d(f^n(\tilde{x}_0), \omega(\tilde{x}, f)) = 0$, there is a j_0 sufficiently large such that

$$
d(f^{n_j}(\tilde{x}_0), \omega(\tilde{x}, f)) < \frac{1}{2}d(\tilde{x}_{\tilde{i}}, \omega(\tilde{x}, f)), \quad \text{for all} \quad j \geq j_0 .
$$

Now, choose $N = n_{j_0} > N_1$. Since $f^{n_{j_0}}(W) \ni \tilde{x}_{\tilde{i}}$, there is an $\hat{x} \in W$ such that $f^{n_{j_0}}(\hat{x}) = \tilde{x}_{\tilde{i}}$. Therefore,

$$
\begin{aligned}
|f^{n_{j_0}}(\tilde{x}_0) - f^{n_{j_0}}(\hat{x})| = |f^{n_{j_0}}(\tilde{x}_0) - \tilde{x}_{\tilde{i}}| &\geq d(\tilde{x}_{\tilde{i}}, \omega(\tilde{x}, f)) - d(f^{n_{j_0}}(\tilde{x}_0), \omega(\tilde{x}, f)) \\
&\geq \frac{1}{2}d(\tilde{x}_{\tilde{i}}, \omega(\tilde{x}, f)) = \delta.
\end{aligned}
$$

Hence, Lemma 10.18 holds for Case (2.b.i).

Finally, consider Case (2.b.ii). Since $f^{n_0}(W) \cap \omega(\tilde{x}, f) \neq \emptyset$, there is an $\hat{x} \in W$ such that $f^{n_0}(\hat{x}) = \tilde{x}_{\tilde{j}}$, for some $\tilde{x}_{\tilde{j}} \in \omega(\tilde{x}, f) \subseteq E$. Pick a point $y_0 \in f^{n_0}(W) \backslash \{f^{n_0}(\hat{x})\}$. Let $\hat{\hat{x}} \in W$ be such that $f^{n_0}(\hat{\hat{x}}) = y_0$. Since $\omega(\tilde{x}, f) \subseteq E$ and E is finite, $\omega(\tilde{x}, f)$ is finite and has, say, k_1 elements. By Lemma 10.15, we have $f^{k_1}(x^*) = x^*$ for all $x^* \in \omega(\tilde{x}, f)$. Define $F(x) = f^{k_1}(x)$. Then each $x^* \in \omega(\tilde{x}, f)$ is a fixed point of F, and F satisfies $F \in PM(I, I)$ and $U(F) = I$ as well, by Proposition 10.12 (v). By Lemma 10.15, there exists a $\delta > 0$ and N_1 (depending on $\tilde{x}_{\tilde{j}}$) such that

$$
|F^{N_1}(y_0) - F^{N_1}(\tilde{x}_{\tilde{j}})| = |F^{N_1}(y_0) - \tilde{x}_{\tilde{j}}| \geq 2\delta .
$$

Let $N = N_1 k_1 + n_0$. Then

$$
|f^N(\hat{\hat{x}}) - f^N(\hat{x})| = |F^{N_1}(y_0) - F^{N_1}(\tilde{x}_{\tilde{j}})| \geq 2\delta .
$$

Hence, by an application of the triangle inequality, we have

$$\text{either } |f^N(\tilde{x}_0) - f^N(\hat{x})| > \delta, \quad \text{or} \quad |f^N(\tilde{x}_0) - f^N(\hat{\hat{x}})| > \delta .$$

Therefore, Lemma 10.18 holds for Case (2.b.ii).

The proof is complete. □

Proof of necessity of Theorem 10.10. Let $E = \{\tilde{x}_0, \tilde{x}_1, \ldots, \tilde{x}_k\}$ be the set of all extremal points of f. By Lemma 10.18, for any interval $W \ni \tilde{x}_i$, there is a $\delta_i > 0$ (independent of W) such that there is an $\hat{x}_i \in W \setminus \{x_i\}$ and N_i (dependent on $\hat{x}_i$) satisfying

$$|f^{N_i}(\hat{x}_i) - f^{N_i}(\tilde{x}_i)| > \delta_i, \qquad i = 0, 1, 2, \ldots, k. \tag{10.25}$$

Define $2\delta \equiv \min\{\delta_i \mid i = 0, 1, \ldots, k\}$. For any $x \in I$ and any interval $W \ni x$, by Lemma 10.17, for some $N' \in \mathbb{Z}^+$, $f^{N'}(W)$ contains an extremal point, say $\tilde{x}_{\bar{j}}$, i.e., $\tilde{x}_{\bar{j}} \in f^{N'}(W)$. Since $f^{N'}(W)$ is an interval with positive length, by (10.25) and Lemma 10.18, there is an $\hat{x} \in f^{N'}(W)$ and an $N_{\bar{j}}$ such that

$$|f^{N_{\bar{j}}}(\hat{x}) - f^{N_{\bar{j}}}(\tilde{x}_{\bar{j}})| \geq \delta_{\bar{j}} \geq 2\delta .$$

Now, let $N = N_{\bar{j}} + N'$, $y_1, y_2 \in W$ satisfy $f^{N'}(y_1) = \hat{x}$, $f^{N'}(y_2) = \tilde{x}_{\bar{j}}$. We have

$$|f^N(y_1) - f^N(y_2)| = |f^{N_{\bar{j}}}(\hat{x}) - f^{N_{\bar{j}}}(\tilde{x}_{\bar{j}})| \geq 2\delta .$$

Therefore, for any $x \in W$, by an application of the triangle inequality, we have

$$\text{either } |f^N(y_1) - f^N(x)| \geq \delta \quad \text{or} \quad |f^N(y_2) - f^N(x)| \geq \delta .$$

The sensitive dependence of f on initial data has been proven. □

10.5   PROPERTIES OF THE SET $E(f)$

Recall from Definition 10.1 part (3) that

$$E(f) = \{x \in I \mid \gamma(x, f) > 0\}.$$

Denote

$$\gamma(f) = \limsup_{n \to \infty} \frac{1}{n} \ln V_I(f^n).$$

Theorem 10.19   Let $f \in PM(I, I)$. Then

$$\gamma(f) = \sup_{x \in I} \gamma(x, f),$$

and there exists an $x \in I$ such that $\gamma(f) = \gamma(x, f)$.

Proof. We assume that $f\colon I \to I$ is onto. It is clear that $\gamma(x, f)$ is no greater than $\limsup_{n\to\infty} \frac{1}{n} \ln V_I(f^n)$ for each $x \in I$. Choose the two subintervals I_{11}, I_{12} of I such that their lengths satisfy $|I_{11}| = |I_{12}| = \frac{1}{2}|I|$ and $I = I_{11} \cup I_{12}$. Then,

$$\limsup_{n\to\infty} \frac{1}{n} \ln \left(V_{I_{11}}(f^n) + V_{I_{12}}(f^n)\right) = \limsup_{n\to\infty} \frac{1}{n} \ln V_I(f^n) = \gamma(f) .$$

This indicates that

$$\limsup_{n\to\infty} \frac{1}{n} \ln V_{I_{1i}}(f^n) = \gamma(f) \quad \text{for} \quad i = 1 \quad \text{or} \quad i = 2.$$

Denote this interval by I_1. Similarly, we can find a subinterval I_2 of I_1 whose length $|I_2|$ equals $\frac{1}{2}|I_1|$ such that

$$\limsup_{n\to\infty} \frac{1}{n} \ln V_{I_2}(f^n) = \gamma(f).$$

Inductively, we can find a decreasing sequence of closed intervals $I_1 \supset I_2 \supset \cdots \supset I_n \supset \cdots$ whose lengths $|I_n| \to 0$ as $n \to \infty$. Thus, $\bigcap_{n\geq 1} I_n$ contains a single point x_0, whose fluctuations satisfy $\gamma(x_0, f) = \gamma(f)$. □

We know from Theorem 10.19 that f has rapid fluctuations if and only if $E(f) \neq \emptyset$. This is a characterization of the chaotic behavior of the I3DS in (10.1) in terms of the map f.

Theorem 10.20 Let $f \in PM(I, I)$. Then the map $\gamma(\cdot, f)\colon I \to [0, +\infty]$ is upper semi-continuous.

Proof. For a point $x_0 \in I$ and a number $\varepsilon > 0$, since $\gamma(x_0, f) = \lim_{\delta\to 0} \gamma([x_0 - \delta, x_0 + \delta] \cap I, f)$, we can find some $\delta > 0$ such that

$$\limsup_{n\to\infty} \frac{1}{n} \ln V_{[x_0-\delta, x_0+\delta]\cap I}(f^n) < \gamma(x_0, f) + \varepsilon .$$

Therefore, $\gamma(x, f) < \gamma(x_0, f) + \varepsilon$ for every x in $(x_0 - \delta, x_0 + \delta) \cap I$. This is just the definition of upper semi-continuity of x_0. □

From Theorem 10.20, we have the following.

Corollary 10.21 Let $f \in PM(I, I)$. We have that $\gamma(x, f) \leq \gamma(z, f)$ for every z in the closure $\overline{\{f^n(x)\colon n \geq 0\}}$ of the orbit $\mathrm{orb}(x, f) = \{f^n(x)\colon n \geq 0\}$. Thus, $\overline{\{f^n(x)\colon n \geq 0\}} \subset E(f)$ if $x \in E(f)$. □

The following gives the conditions under which the function $\gamma(\cdot, f)$ is almost constant.

Theorem 10.22 Let $f \in PM(I, I)$. Then

(1) If f is topologically mixing, then $\gamma(x, f) = \text{const.} > 0$, for all $x \in \text{int } I$;

(2) If f is topologically transitive, then

$$\gamma(x, f) = \text{const.} > 0, \qquad \forall x \in \text{int } I - \{p\},$$

for some fixed point $p \in I$. □

Exercise 10.23   Prove Theorem 10.22. □

Exercise 10.24   Let $f \in C(I, I)$, piecewise monotone with finitely many critical points. Show that $B(f)$ is open in I and $B(f) \cup U(f) = I$. □

Exercise 10.25   Prove (3) in Theorem 10.2. □

Exercise 10.26   Consider the quadratic map

$$f(x) = \mu x(1 - x).$$

Show that

(i) If $1 < \mu < 3$, then $B(f) = [0, 1]$.

(ii) If $\mu > 3$, then $0 \in U(f)$.

(iii) Find a condition on μ such that $E(f) \neq \emptyset$. □

Exercise 10.27   Let $f : [0, 1] \to [0, 1]$ be defined as

$$f(x) = \begin{cases} 3x & 0 \leq x \leq \dfrac{1}{3} \\[2mm] 1 & \dfrac{1}{3} < x \leq \dfrac{2}{3} \\[2mm] -3x + 3 & \dfrac{2}{3} < x \leq 1. \end{cases}$$

Show that

(i) For any $x \in (\frac{1}{3}, \frac{2}{3})$, we have $x \in B(f)$.

(ii) Furthermore, for any $x \in [0, 1] - \mathcal{C}, x \in B(f)$.

(iii) $E(f) = \mathcal{C}$.

Here $\mathcal{C}$ denotes the classical Cantor set. □

NOTES FOR CHAPTER 10

Chapter 9 studies maps from a more functional-analytic and function-space point of view, utilizing the notion of rapid fluctuations in Chapter 8. In this approach, we are essentially dealing with maps on an infinite-dimensional space. Such dynamical systems are called *I3DS*, which occur naturally in the study of delay equations or even PDEs.

In particular, our Sections 10.2–10.3 are from [38] and Theorem 10.10 in Section 10.4 is from [17].

Introduction to Continuous-Time Dynamical Systems

A nonlinear differential equation or system can be described in the form

$$\begin{cases} \dot{x}(t) = f(x(t), t), & t_0 \le t < T, \\ x(t_0) = x_0 \in \mathbb{R}^N, \end{cases} \tag{A.1}$$

where

$$f: \ \mathbb{R}^N \times \mathbb{R} \to \mathbb{R}^N \tag{A.2}$$

is a sufficiently smooth function; $t_0, T \in \mathbb{R}$ and $x_0 \in \mathbb{R}^N$ are given. The state of x at time t_0, x_0, is called the *initial condition*. If f in (A.1) and (A.2) is independent of t, we say that (A.1) is an *autonomous* system. (Otherwise, the system (A.1) is called nonautonomous, time-dependent or time-varying.) Assume that the solution of (A.1) exists for any $T > t_0$. One is interested in studying the *asymptotic behavior* of (A.1) when $T \to \infty$. This study for continuous-time systems can obviously be extremely challenging, if not more so than the discrete-time cases that we have been studying in the preceding chapters. Nevertheless, as we will see in this Appendix A, one can deduce important information about the asymptotic behavior of (A.1) through the use of the *Poincaré sections*. The main objective of this Appendix A is to give the readers just a little flavor of continuous-time dynamical systems, as an in-depth account would require a large tome to do the job.

A.1 THE LOCAL BEHAVIOR OF 2-DIMENSIONAL NONLINEAR SYSTEMS

We consider mainly autonomous systems (A.1), i.e., $f(x, t) \equiv f(x)$ for all x, t. As we have seen in Definition A.1, the local behavior of being attracting, repelling or neutral for fixed or periodic points is critical in the analysis. This is analyzed through *linearization*.

We first introduce the following.

Definition A.1   A point $x_0 \in \mathbb{R}^N$ is called an *equilibrium point* of an autonomous equation $\dot{x}(t) = f(x(t))$ if $f(x_0) = 0$. ☐

If $\boldsymbol{x}_0 \in \mathbb{R}^N$ is an equilibrium point of (A.1) and $\widetilde{\boldsymbol{x}}_0 \in \mathbb{R}^N$ satisfies $|\boldsymbol{x}_0 - \widetilde{\boldsymbol{x}}_0| < \delta$ for some very small $\delta > 0$, then the solution $\widetilde{\boldsymbol{x}}(t)$ of

$$\begin{cases} \dot{\widetilde{\boldsymbol{x}}}(t) = \boldsymbol{f}(\widetilde{\boldsymbol{x}}(t)), & t > t_0, \\ \widetilde{\boldsymbol{x}}(t_0) = \widetilde{\boldsymbol{x}}_0 \in \mathbb{R}^N, \end{cases}$$

satisfies

$$|\widetilde{\boldsymbol{x}}(t) - \boldsymbol{x}(t)| < \delta', \qquad t_0 \le t \le t_1,$$

for some δ' (depending on δ) for some $t_1 > t_0$. Thus,

$$\boldsymbol{\eta}(t) \equiv \widetilde{\boldsymbol{x}}(t) - \boldsymbol{x}_0$$

satisfies

$$\dot{\boldsymbol{\eta}}(t) = \dot{\widetilde{\boldsymbol{x}}}(t) = \boldsymbol{f}(\widetilde{\boldsymbol{x}}(t)) - \boldsymbol{f}(\boldsymbol{x}_0) = D_{\boldsymbol{x}} f(\boldsymbol{x}_0) \cdot \boldsymbol{\eta}(t) + \mathcal{O}(|\boldsymbol{\eta}(t)|^2).$$

Thus, for $\boldsymbol{\eta}(t)$ small, the first order term

$$\dot{\boldsymbol{\eta}}(t) \approx D_{\boldsymbol{x}} f(\boldsymbol{x}_0)\boldsymbol{\eta}(t) \tag{A.3}$$

dominates in (A.3). We call

$$\dot{\boldsymbol{y}}(t) = A\boldsymbol{y}(t), \qquad A \equiv D_{\boldsymbol{x}} f(\boldsymbol{x}_0), \tag{A.4}$$

the *linearized equation* of $\dot{\boldsymbol{x}}(t) = \boldsymbol{f}(\boldsymbol{x}(t))$ at $\boldsymbol{x}_0$.

We may now utilize a standard procedure in linear algebra by converting the $N \times N$ constant matrix A into the *Jordan canonical form* by finding a (similarity) matrix S such that

$$\boldsymbol{z}(t) = S\boldsymbol{y}(t).$$

Then

$$\dot{\boldsymbol{z}}(t) = S\dot{\boldsymbol{y}}(t) = SA\boldsymbol{y}(t) = SAS^{-1}\boldsymbol{z}(t) \equiv J\boldsymbol{z}(t),$$

where

$$J = \begin{bmatrix} J_1 & 0 & \cdots & 0 \\ 0 & J_2 & \ddots & \vdots \\ \vdots & \ddots & \ddots & 0 \\ 0 & \cdots & 0 & J_k \end{bmatrix}$$

with J_ℓ, $1 \le \ell \le k$, taking one of the following two forms:

$$J_\ell = \begin{bmatrix} \lambda_\ell & 0 & \cdots & 0 \\ 0 & \lambda_\ell & \ddots & \vdots \\ \vdots & \ddots & \ddots & 0 \\ 0 & \cdots & 0 & \lambda_\ell \end{bmatrix} \quad \text{or} \quad J_\ell = \begin{bmatrix} \lambda_\ell & 1 & & 0 \\ 0 & \lambda_\ell & \ddots & \\ \vdots & \ddots & \ddots & 1 \\ 0 & \cdots & 0 & \lambda_\ell \end{bmatrix}.$$

The case $N = 2$, i.e., two-dimensional autonomous systems, is the easiest to understand and to visualize (besides the somewhat trivial case $N = 1$) as well as offers significant clues to more complicated cases for systems in higher dimensional spaces.

Thus, we consider a real 2×2 matrix A. We have the following possibilities for J:

$$\text{Case (1)} \quad J = \begin{bmatrix} \lambda_1 & 0 \\ 0 & \lambda_2 \end{bmatrix}, \quad \lambda_1, \lambda_2 \in \mathbb{R}; \tag{A.5}$$

$$\text{Case (2)} \quad J = \begin{bmatrix} \lambda & 1 \\ 0 & \lambda \end{bmatrix}, \quad \lambda \in \mathbb{R}; \tag{A.6}$$

$$\text{Case (3)} \quad J = \begin{bmatrix} \alpha + i\beta & 0 \\ 0 & \alpha - i\beta \end{bmatrix}, \quad \text{or, equivalently,}$$

$$J = \begin{bmatrix} \alpha & -\beta \\ \beta & \alpha \end{bmatrix}; \quad \alpha, \beta \in \mathbb{R}. \tag{A.7}$$

Case (1) may be further subdivided into the following:

(1.i) $\lambda_1 < \lambda_2 < 0$ $\Big\}$ stable node;
(1.ii) $\lambda_2 < \lambda_1 < 0$
(1.iii) $\lambda_2 > \lambda_1 > 0$ $\Big\}$ unstable node;
(1.iv) $\lambda_1 > \lambda_2 > 0$
(1.v) $\lambda_1 = \lambda_2 > 0$ $\big\}$ unstable star node;
(1.vi) $\lambda_1 = \lambda_2 < 0$ stable star node;
(1.vii) $\lambda_1 = \lambda_2 = 0$
(1.viii) $\lambda_2 = 0, \lambda \gtreqless 0$ $\Big\}$ the equilibrium points may be nonisolated;
(1.ix) $\lambda_2 = 0, \lambda_1 \gtreqless 0$
(1.x) $\lambda_1 > 0 > \lambda_2$ $\big\}$ saddle points.
(1.xi) $\lambda_2 > 0 > \lambda_1$

We discuss them in separate groups below.

SUBCASES (1.I)–(1.VI)

$$\begin{bmatrix} \dot{x} \\ \dot{y} \end{bmatrix} = \begin{bmatrix} \lambda_1 & 0 \\ 0 & \lambda_2 \end{bmatrix} \begin{bmatrix} x \\ y \end{bmatrix}, \quad \begin{bmatrix} x(0) \\ y(0) \end{bmatrix} = \begin{bmatrix} x_0 \\ y_0 \end{bmatrix} \tag{A.8}$$

$$\frac{dy/dt}{dx/dt} = \frac{dy}{dx} = \frac{\lambda_2 y}{\lambda_1 x} = \gamma \left(\frac{y}{x} \right), \quad \gamma \equiv \frac{\lambda_2}{\lambda_1}$$

$$\int \frac{dy}{y} = \gamma \int \frac{dx}{x}$$

$$\ln \left| \frac{y}{y_0} \right| = \gamma \ln \left| \frac{x}{x_0} \right|$$

$$|y| = |y_0| \left| \frac{x}{x_0} \right|^\gamma. \tag{A.9}$$

Also from (A.8),

$$x(t) = x_0 e^{\lambda_1 t}, \quad y(t) = y_0 e^{\lambda_2 t}. \tag{A.10}$$

Solutions to Subcases (I.vii)–(I.xi) can be similarly determined. The trajectories of solutions $(x(t), y(t))$ plotted on the (x, y)-plane are called *phase portraits*. We illustrate them through various examples and graphics in this section.

Example A.2   (A stable node (Subcases (1.i) and (1.ii)))   For the differential equation

$$\frac{d}{dt}\begin{bmatrix} x(t) \\ y(t) \end{bmatrix} = \begin{bmatrix} -3 & -1 \\ 2 & 0 \end{bmatrix}\begin{bmatrix} x(t) \\ y(t) \end{bmatrix} \equiv A\begin{bmatrix} x(t) \\ y(t) \end{bmatrix},$$

we have

$$A = SJS^{-1} \equiv \begin{bmatrix} -1 & 1 \\ 2 & -1 \end{bmatrix}\begin{bmatrix} -1 & 0 \\ 0 & -2 \end{bmatrix}\begin{bmatrix} 1 & 1 \\ 2 & 1 \end{bmatrix}, \text{ with } J = \begin{bmatrix} -1 & 0 \\ 0 & -2 \end{bmatrix},$$

thus $\lambda_2 = -2 < \lambda_1 = -1 < 0$. So the equilibrium point $(x, y) = (0, 0)$ is a stable node. See Fig. A.1 for the phase portrait.   □

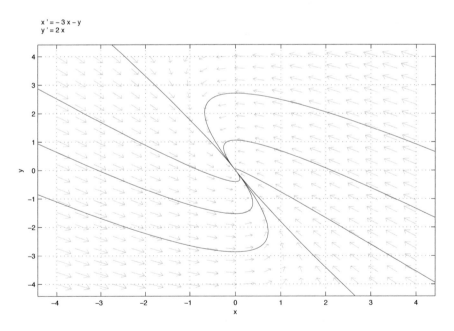

Figure A.1: The phase portrait for Example A.2. This is a stable node (Subcase (1.i) and (1.ii)).

Example A.3   (An unstable node (Subcases (1.iii) and (1.iv)))   For the differential equation

$$\frac{d}{dt}\begin{bmatrix} x(t) \\ y(t) \end{bmatrix} = \begin{bmatrix} 0 & -1 \\ 2 & 3 \end{bmatrix}\begin{bmatrix} x(t) \\ y(t) \end{bmatrix} \equiv A\begin{bmatrix} x(t) \\ y(t) \end{bmatrix},$$

we have

$$A = SJS^{-1} = \begin{bmatrix} 1 & -1 \\ -1 & 2 \end{bmatrix} = \begin{bmatrix} 1 & 0 \\ 0 & 2 \end{bmatrix} \begin{bmatrix} 2 & 1 \\ 1 & 1 \end{bmatrix}, \text{ with } J \begin{bmatrix} 1 & 0 \\ 0 & 2 \end{bmatrix},$$

thus $\lambda_2 = 2 > \lambda_1 \equiv 1 > 0$. So the equilibrium point $(x, y) = (0, 0)$ is an unstable node. See Fig. A.2 for the phase portrait. $\square$

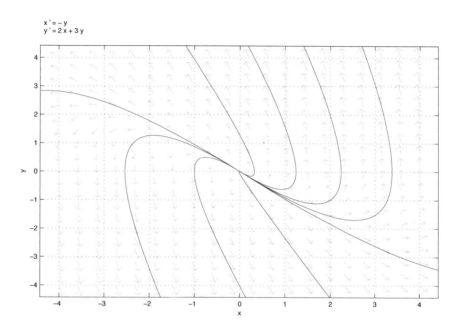

x' = -y
y' = 2 x + 3 y

Figure A.2: The phase portrait for Example A.3. This is an unstable node. (Subcases (1.i) and (1.ii)).

Example A.4   (An unstable star node (Subcase (1.v)))   The phase portrait of the differential equation

$$\frac{d}{dt}\begin{bmatrix} x(t) \\ y(t) \end{bmatrix} = \begin{bmatrix} 3 & 0 \\ 0 & 3 \end{bmatrix} \begin{bmatrix} x(t) \\ y(t) \end{bmatrix}$$

is illustrated in Fig. A.3. We have $\lambda_1 = \lambda_2 = 3 > 0$, and thus an unstable star node. $\square$

Example A.5   (A stable star node (Subcase (1.vi)))   The phase portrait of the differential equation

$$\frac{d}{dt}\begin{bmatrix} x(t) \\ y(t) \end{bmatrix} = \begin{bmatrix} -1 & 0 \\ 0 & -1 \end{bmatrix} \begin{bmatrix} x(t) \\ y(t) \end{bmatrix}$$

is illustrated in Fig. A.4. We have $\lambda_1 = \lambda_2 = -1 < 0$, and thus a stable star node. $\square$

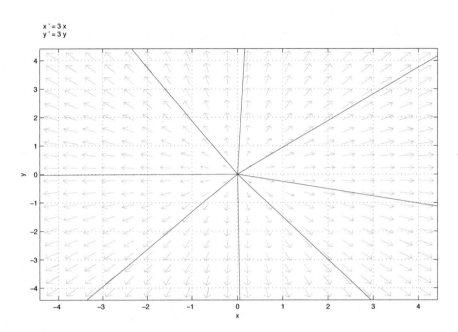

Figure A.3: The phase portrait for Example A.4. This is an unstable star node. (Subcase (1.v)).

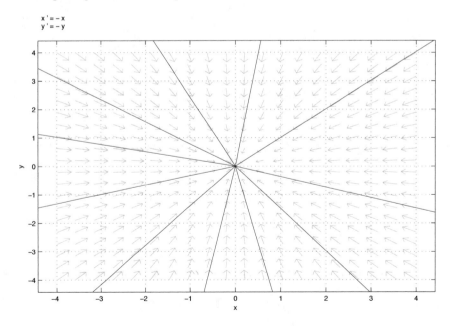

Figure A.4: The phase portrait for Example A.5. This is a stable star node (Subcase (1.vi)).

Example A.6   (Non-isolated equilibrium points (Subcases (1.vii), (1.viii), (1.ix))) Consider

$$
\text{(a)} \quad \frac{d}{dt}\begin{bmatrix} x(t) \\ y(t) \end{bmatrix} = \begin{bmatrix} -2 & 2 \\ -4 & 4 \end{bmatrix}\begin{bmatrix} x(t) \\ y(t) \end{bmatrix} \equiv A\begin{bmatrix} x(t) \\ y(t) \end{bmatrix}, \quad \text{and} \tag{A.11}
$$

$$
\text{(b)} \quad \frac{d}{dt}\begin{bmatrix} x(t) \\ y(t) \end{bmatrix} = \begin{bmatrix} 2 & -2 \\ 4 & -4 \end{bmatrix}\begin{bmatrix} x(t) \\ y(t) \end{bmatrix} = -A\begin{bmatrix} x(t) \\ y(t) \end{bmatrix}. \tag{A.12}
$$

Since

$$
A = SJS^{-1} = \begin{bmatrix} 1 & 1 \\ 2 & 1 \end{bmatrix}\begin{bmatrix} 2 & 0 \\ 0 & 0 \end{bmatrix}\begin{bmatrix} -1 & 1 \\ 2 & -1 \end{bmatrix}, \quad \text{with } J = \begin{bmatrix} 2 & 0 \\ 0 & 0 \end{bmatrix},
$$

we see that (a) and (b) are examples of case (1.ix). The phase protraits of (a) and (b) are illustrated in Fig. A.5 (a) and (b), respectively. Every trajectory consists of two semiinfinite line segments.
  If

$$
\frac{d}{dt}\begin{bmatrix} x(t) \\ y(t) \end{bmatrix} = \begin{bmatrix} 0 \\ 0 \end{bmatrix}, \quad \text{i.e.,} \quad A \equiv \begin{bmatrix} 0 & 0 \\ 0 & 0 \end{bmatrix},
$$

this becomes case (1.vii). Then every point on the (x, y)-plane is an equilibrium point.   □

Example A.7   (A saddle point (Subcases (1.x) and (1.xi)))   For the differential equation

$$
\frac{d}{dt}\begin{bmatrix} x(t) \\ y(t) \end{bmatrix} = \begin{bmatrix} -4 & 3 \\ -6 & 5 \end{bmatrix}\begin{bmatrix} x(t) \\ y(t) \end{bmatrix} \equiv A\begin{bmatrix} x(t) \\ y(t) \end{bmatrix},
$$

we have

$$
A = SJS^{-1} = \begin{bmatrix} 1 & 1 \\ 2 & 1 \end{bmatrix}\begin{bmatrix} 2 & 0 \\ 0 & -1 \end{bmatrix}\begin{bmatrix} -1 & 1 \\ 2 & -1 \end{bmatrix}, \quad \text{with } J = \begin{bmatrix} 2 & 0 \\ 0 & -1 \end{bmatrix}.
$$

Thus, $\lambda_2 = -1 < 0 < \lambda_1 = 2$. So the equilibrium point $(x, y) = (0, 0)$ is a saddle point. See Fig. A.6 for the phase portrait.   □

Next we consider Case (2), which may be further subdivided into the following:

(2.i) $A = \begin{bmatrix} 0 & 1 \\ 0 & 1 \end{bmatrix}$, i.e., $\lambda = 0$;

(2.ii) $A = \begin{bmatrix} \lambda & 1 \\ 0 & \lambda \end{bmatrix}$, where $\lambda \neq 0$.

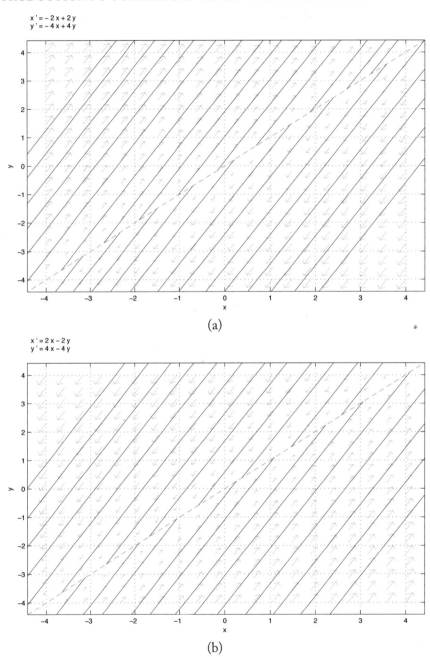

Figure A.5: The phase portrait for Example A.6. (Subcases (1.viii) and (1.ix).) Here (a) and (b) correspond, respectively, to (A.11) and (A.12). Every trajectory consists of two semiinfinite line segments. The equilibrium points consist of an entire line, so they are not isolated.

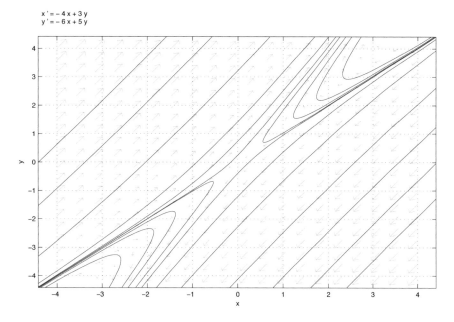

Figure A.6: The phase portrait for Example A.7. This is a saddle point (Subcases (1.x) and (1.xi)). Saddle points are always unstable.

SUBCASE (2.I)

We have $\dot{x} = y$ and $\dot{y} = 0$. Thus,

$$y(t) = y_0 = \text{constant}, x(t) = y_0 t.$$

Example A.8   (Non-isolated equilibrium points (Subcase (2.i)))   Consider the differential equation

$$\frac{d}{dt} \begin{bmatrix} x(t) \\ y(t) \end{bmatrix} = \begin{bmatrix} 2 & -1 \\ 4 & -2 \end{bmatrix} \begin{bmatrix} x(t) \\ y(t) \end{bmatrix} \equiv A \begin{bmatrix} x(t) \\ y(t) \end{bmatrix},$$

where

$$A = SJS^{-1} = \begin{bmatrix} 1 & 1 \\ 2 & 1 \end{bmatrix} \begin{bmatrix} 0 & 1 \\ 0 & 0 \end{bmatrix} \begin{bmatrix} -1 & 1 \\ 2 & -1 \end{bmatrix}, \text{ with } J = \begin{bmatrix} 0 & 1 \\ 0 & 0 \end{bmatrix}.$$

Thus, this belongs to Subcase (2.i). The phase portrait is plotted in Fig. A.7. All points on the line $2x - y = 0$ are equilibrium points.   □

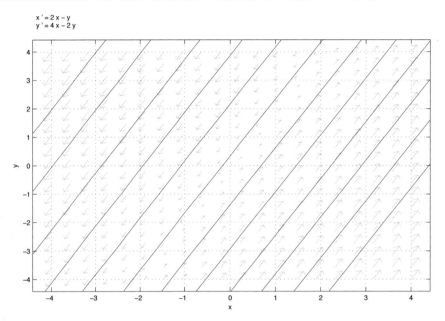

x' = 2 x − y
y' = 4 x − 2 y

Figure A.7: Phase portrait for Example A.8, where all points on the line $2x - y = 0$ are equilibrium points. The other trajectories are parallel lines (Subcase (2.i)).

SUBCASE (2.II)

We have $\dot{x} = \lambda x + y$ and $\dot{y} = \lambda y$. Thus,

$$y(t) = y_0 e^{\lambda t} \quad \text{and} \quad x(t) = x_0 e^{\lambda t} + y_0 t e^{\lambda t},$$

$$t = \frac{1}{\lambda} \ln \left| \frac{y}{y_0} \right|,$$

$$x = \left(\frac{x_0}{y_0} + \frac{1}{\lambda} \ln \left| \frac{y}{y_0} \right| \right) y.$$

Example A.9 (Improper stable or unstable node (Subcase (2.ii))) Consider

$$\text{(a)} \quad \frac{d}{dt} \begin{bmatrix} x(t) \\ y(t) \end{bmatrix} = \begin{bmatrix} 0 & -1 \\ 4 & -4 \end{bmatrix} \begin{bmatrix} x(t) \\ y(t) \end{bmatrix} \equiv A_1 \begin{bmatrix} x(t) \\ y(t) \end{bmatrix}, \tag{A.13}$$

where

$$A_1 = S J_1 S^{-1} = \begin{bmatrix} 1 & 1 \\ 2 & 1 \end{bmatrix} \begin{bmatrix} -2 & 1 \\ 0 & -2 \end{bmatrix} \begin{bmatrix} -1 & 1 \\ 2 & -1 \end{bmatrix}, \text{ with } J_1 = \begin{bmatrix} -2 & 1 \\ 0 & -2 \end{bmatrix};$$

and

$$\text{(b)} \quad \frac{d}{dt} \begin{bmatrix} x(t) \\ y(t) \end{bmatrix} = \begin{bmatrix} 4 & -1 \\ 4 & 0 \end{bmatrix} \begin{bmatrix} x(t) \\ y(t) \end{bmatrix} \equiv A_2 \begin{bmatrix} x(t) \\ y(t) \end{bmatrix}, \tag{A.14}$$

where

$$A_2 = S J_2 S^{-1} = \begin{bmatrix} 1 & 1 \\ 2 & 1 \end{bmatrix} \begin{bmatrix} 2 & 1 \\ 0 & 2 \end{bmatrix} \begin{bmatrix} -1 & 1 \\ 2 & -1 \end{bmatrix}, \text{ with } J_2 = \begin{bmatrix} 2 & 1 \\ 0 & 2 \end{bmatrix}.$$

The phase portraits for (A.13) and (A.14) are given in Fig. A.8. The equilibrium point $(x, y) = (0, 0)$ is, respectively, a stable and an unstable improper node.   □

Finally, we treat Case (3) by writing $A = \begin{bmatrix} \alpha & -\beta \\ \beta & \alpha \end{bmatrix}$. We have

$$\begin{cases} \dot{x} = \alpha x - \beta y, \\ \dot{y} = \beta x + \alpha y. \end{cases} \tag{A.15}$$

Using polar coordinates

$$x(t) = r(t) \cos(\theta(t)), \qquad y(t) = r(t) \sin(\theta(t)),$$

we obtain

$$\begin{cases} \dot{r} \cos \theta - r \dot{\theta} \sin \theta = \alpha r \cos \theta - \beta r \sin \theta, \\ \dot{r} \sin \theta - r \dot{\theta} \cos \theta = \beta r \cos \theta + \alpha r \sin \theta. \end{cases}$$

Hence,

$$\begin{cases} \dot{r} = \alpha r, \\ r \dot{\theta} = \beta r, \end{cases} \text{ implying } \begin{cases} r(t) = r_0 e^{\alpha t}, \\ \theta(t) = \beta t. \end{cases} \tag{A.16}$$

We see that the stability of the equilibrium point is completely determined by the sign of α.

Example A.10   (Stable and unstable spirals (Case (3), $\alpha \neq 0$))   Consider

$$\text{(a)} \quad \frac{d}{dt} \begin{bmatrix} x(t) \\ y(t) \end{bmatrix} = \begin{bmatrix} -3 & -2 \\ 2 & -3 \end{bmatrix} \begin{bmatrix} x(t) \\ y(t) \end{bmatrix}, \tag{A.17}$$

and

$$\text{(b)} \quad \frac{d}{dt} \begin{bmatrix} x(t) \\ y(t) \end{bmatrix} = \begin{bmatrix} 3 & 2 \\ -2 & 3 \end{bmatrix} \begin{bmatrix} x(t) \\ y(t) \end{bmatrix}. \tag{A.18}$$

For (A.17), we have $\alpha = -3 < 0$. Therefore, the equilibrium point $(x, y) = (0, 0)$ is stable. All trajectories spiral toward the origin; see Fig. A.9(a).

For (A.18), we have $\alpha = 3 > 0$. Therefore, the equilibrium point $(x, y) = (0, 0)$ is unstable. All trajectories spiral away from the origin; see Fig. A.9(b).   □

Example A.11   (Center (Case (3), $\alpha = 0$))   Consider

$$\frac{d}{dt} \begin{bmatrix} x(t) \\ y(t) \end{bmatrix} = \begin{bmatrix} 0 & -2 \\ 2 & 0 \end{bmatrix} \begin{bmatrix} x(t) \\ y(t) \end{bmatrix}.$$

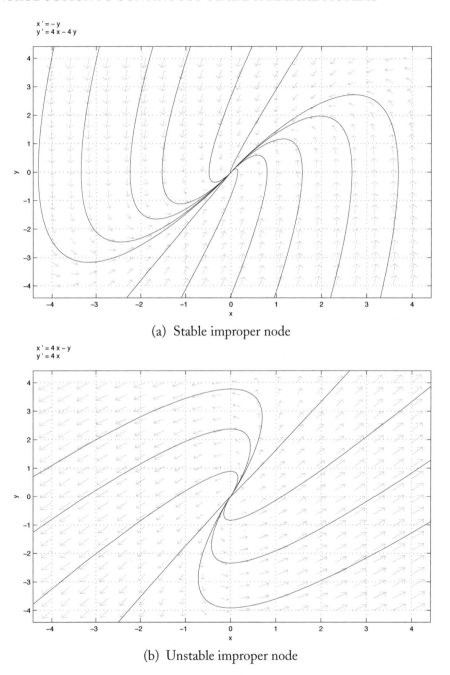

(a) Stable improper node

(b) Unstable improper node

Figure A.8: The phase portraits for equations (A.13) and (A.14) are, respectively, stable and unstable improper nodes (Subcase (2.ii)).

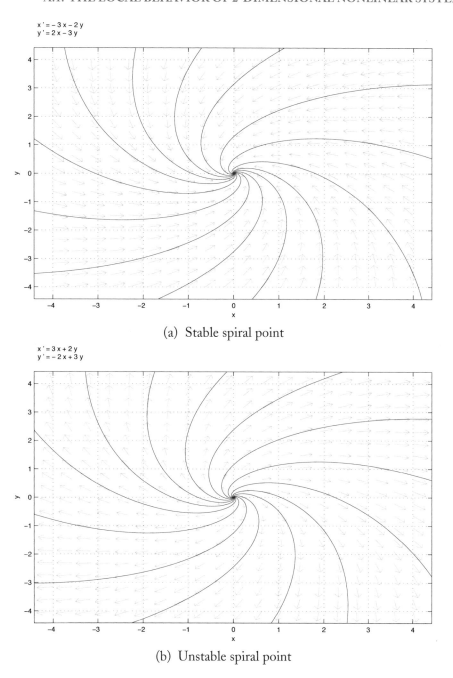

(a) Stable spiral point

(b) Unstable spiral point

Figure A.9: Phase portraits for equations (A.17) and (A.18) in Example A.10, as shown in, respectively, (a) and (b). They correspond to Case (3), with $\alpha \neq 0$.

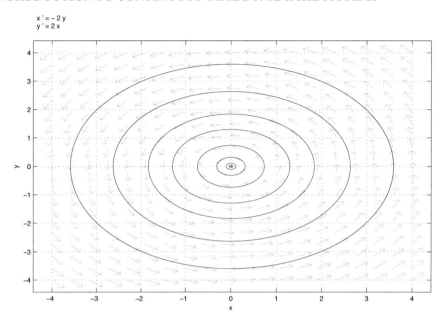

Figure A.10: The phase portrait for Example A.11. This equilibrium point is a center (Case (3), $\alpha = 0$).

Here we see that $\alpha = 0$ in (A.15). The phase portrait consist of circles or ellipses (which are periodic solutions) enclosing the equilibrium point $(x, y) = (0, 0)$. □

In summary, we have a total of nine types of phase portraits:

$$
\left.\begin{array}{r}
\text{parallel lines} \\
\text{all } \mathbb{R}^2 \\
\text{centers}
\end{array}\right\} \text{ these 3 types have neutral stability}
$$

$$
\left.\begin{array}{r}
\text{nodes} \\
\text{stars} \\
\text{saddle points} \\
\text{improper nodes} \\
\text{spirals} \\
\text{two semiinfinite lines}
\end{array}\right\} \text{ these 5 types are either stable or unstable}
$$

We end this section by including the following example, which shows how to analyze and visualize local behaviors of more complicated 2×2 autonomous differential equations.

Example A.12 Consider

$$
\frac{d}{dt}\begin{bmatrix} x(t) \\ y(t) \end{bmatrix} = \begin{bmatrix} x(t) + y(t) \\ -2x(t) - y(t) - x^2(t) \end{bmatrix}. \tag{A.19}
$$

Equilibrium points are determined by points (x, y) satisfying

$$x + y = 0, \quad -2x - y - x^2 = 0.$$

We obtain two equilibria:

$$(0, 0), (-1, 1).$$

The phase portrait of (A.19) is given in Fig. A.11. □

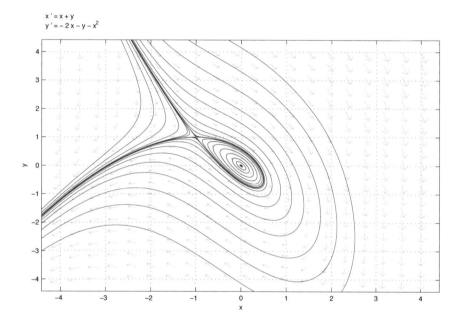

Figure A.11: The phase portrait for Example A.12. Two equilibrium points are $(0,0)$ and $(-1, 1)$. One can see that $(0,0)$ is a center, while $(-1, 1)$ is a saddle point.

Exercise A.13   By linearization, determine the linearized system at $(0,0)$ and $(-1, 1)$ of (A.19) and show that these equilibria are, respectively, center and saddle points. □

A.2   INDEX FOR TWO-DIMENSIONAL SYSTEMS

A powerful method to study two-dimensional systems is the *index theory*. Consider

$$\begin{bmatrix} \dot{x} \\ \dot{y} \end{bmatrix} = \begin{bmatrix} f(x, y) \\ g(x, y) \end{bmatrix} \equiv \boldsymbol{V}(x, y), \qquad (x, y) \in \mathbb{R}^2, \tag{A.20}$$

where $V(x, y)$ denotes the vector field at (x, y). The the angle formed by $V(x, y)$ with the x-axis is

$$\phi = \tan^{-1}\left(\frac{g(x, y)}{f(x, y)}\right), \qquad -\pi < \phi < \pi,$$

where we can determine whether $-\pi < \phi < -\frac{\pi}{2}$, $-\frac{\pi}{2} < \phi < 0$, $0 < \phi < \frac{\pi}{2}$ or $\frac{\pi}{2} < \phi < \pi$ by checking which quadrant $V(x, y)$ points t_0. Thus,

$$\frac{d}{dt}\phi = \frac{\frac{d}{dt}(g/f)}{1 + (g/f)^2} = \frac{(fg' - gf')/f^2}{(g^2 + f^2)/f^2} = \frac{fg' - gf'}{f^2 + g^2}. \tag{A.21}$$

Let Γ be a simple closed curve in $\mathbb{R}^2$. We integrate $\frac{d\phi}{dt}$ along Γ a full circuit. The outcome will be an integral multiple of 2π, assuming that Γ does not contain any equilibrium point of (A.20) (which causes singularity in the denominator of (A.21)). Thus, we define

$$\text{ind}(\Gamma) \equiv \text{index of the curve } \Gamma \equiv \frac{1}{2\pi}\oint d\phi$$
$$= \frac{1}{2\pi}\oint \frac{fg' - gf'}{f^2 + g^2}dt = \frac{1}{2\pi}\oint \frac{fdg - gdf}{f^2 + g^2}.$$

In complex variable theory, the index of a curve is also called the *winding number*.

Theorem A.14   Let Γ be a simple closed curve in $\mathbb{R}^2$ passing through no equilibrium points of (A.20).

 (i) If Γ encloses (in its interior) a single node, star, improper node, center, or spiral point, then $\text{ind}(\Gamma) = 1$;

 (ii) If Γ encloses a single saddle point, then $\text{ind}(\Gamma) = -1$;

 (iii) If Γ itself is a closed orbit of (A.20) (and thus represents a periodic solution), then $\text{ind}(\Gamma) = 1$;

 (iv) If Γ does not enclose any equilibrium point of (A.20), then $\text{ind}(\Gamma) = 0$;

 (v) $\text{ind}(\Gamma)$ is equal to the sum of the indices of all the equilibrium points enclosed within Γ.   □

Instead of giving a proof of the above, we just visualize the properties stated in Theorem A.14 through some illustrations.

Example A.15   The index of a star is $+1$ regardless of its stability or instability. See Figs. A.3 and A.4.   □

Example A.16   The index of a saddle point is -1. See Fig. A.6.   □

Example A.17   Consider the Duffing oscillator

$$\ddot{x} - x + x^3 = 0. \tag{A.22}$$

Its first order form is

$$\begin{bmatrix} \dot{x} \\ y \end{bmatrix} = \begin{bmatrix} y \\ x - x^3 \end{bmatrix} = \boldsymbol{V}(x, y). \tag{A.23}$$

The equation has three equilibria:

$$(0, 0), (1, 0), (-1, 0),$$

where $(0, 0)$ is a saddle point while $(1, 0)$ and $(-1, 0)$ are centers; see Fig. A.12.   □

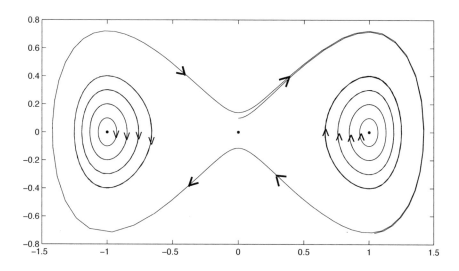

Figure A.12: Phase portrait for the Duffing oscillator in Example A.17, where $(0,0)$, $(1,0)$ and $(-1, 0)$ are the three equilibrium points.

A.3   THE POINCARÉ MAP FOR A PERIODIC ORBIT IN $\mathbb{R}^N$

We now consider a general N-dimensional autonomous system

$$\begin{cases} \dot{\boldsymbol{x}}(t) = \boldsymbol{f}(\boldsymbol{x}(t)), & t > 0, \\ \boldsymbol{x}(0) = \boldsymbol{x}_0 \in \mathbb{R}^N. \end{cases} \tag{A.24}$$

The local property of (A.24) near a nonequilibrium point $\boldsymbol{r} \in \mathbb{R}^N$ (i.e., $\boldsymbol{f}(\boldsymbol{r}) \neq \boldsymbol{0}$) can be nicely described by the following.

Theorem A.18   (The Flow Box Theorem [54, Section V.C]) Let $\mathcal{O} \subseteq \mathbb{R}^N$ be an open neighborhood of a point $\boldsymbol{x}^0$ such that $\boldsymbol{f}(\boldsymbol{x}^0) \neq \boldsymbol{0}$. Then there exists a local coordinate transformation

$y = \psi(x^0)$ near x^0 such that (A.24), with respect to the new coordinates y, is transformed to

$$
\left\{ \quad \dot{y}(t) = \begin{bmatrix} 1 \\ 0 \\ 0 \\ \vdots \\ 0 \end{bmatrix}, \quad t > 0, \quad y(0) = x^0.
\right.
\tag{A.25}
$$

Proof. Since x^0 is not an equilibrium point for the differential equation $\dot{x} = f(x)$,

$$
f\left(x^0\right) = \begin{bmatrix} a_1 \\ a_2 \\ \vdots \\ a_N \end{bmatrix} \neq \begin{bmatrix} 0 \\ 0 \\ \vdots \\ 0 \end{bmatrix}.
$$

We can easily change the variables such that $a_1 \neq 0$. Let $\phi(x, t)$ be the solution flow of the differential equation (A.24). Define the coordinate transformation

$$
x \leftrightarrow y, \quad x = \phi(0, y_2, \ldots, y_N, y_1) \quad \text{where} \quad y = \begin{bmatrix} y_1 \\ y_2 \\ \vdots \\ y_N \end{bmatrix}.
\tag{A.26}
$$

Note that y_1 above corresponds to the t variable: $y_1 = t$. Thus,

$$
\dot{y}_1 = 1.
\tag{A.27}
$$

We now show that near $x = x^0$, the transformation (1) is 1-1. We know that

$$
\phi(y, 0) = y,
$$

and, thus,

$$D_y\phi(y, 0)|_{y=x^0} = D_y y = I_N = \begin{bmatrix} \dfrac{\partial\phi_1}{\partial y_1} & \dfrac{\partial\phi_1}{\partial y_2} & \cdots & \dfrac{\partial\phi_1}{\partial y_N} \\ \vdots & \vdots & & \vdots \\ \vdots & \vdots & & \vdots \\ \dfrac{\partial\phi_N}{\partial y_1} & \dfrac{\partial\phi_N}{\partial y_2} & \cdots & \dfrac{\partial\phi_N}{\partial y_N} \end{bmatrix}$$

$$= \begin{bmatrix} \dfrac{\partial\phi_1}{\partial y_1} & \dfrac{\partial\phi_1}{\partial y_2} & \cdots & \dfrac{\partial\phi_1}{\partial y_N} \\ \vdots & & & \\ \vdots & & I_{N-1} & \\ \dfrac{\partial\phi_N}{\partial y_1} & & & \end{bmatrix}, \tag{A.28}$$

where I_{n-1} is the $(N-1) \times (N-1)$ identity matrix. Now

$$D_y\phi(0, y_2, \ldots, y_N, y_1) = \begin{bmatrix} \dfrac{\partial\phi_1}{\partial t} & \dfrac{\partial\phi_1}{\partial y_2} & \cdots & \dfrac{\partial\phi_1}{\partial y_N} \\ \dfrac{\partial\phi_2}{\partial t} & \dfrac{\partial\phi_2}{\partial y_2} & & \dfrac{\partial\phi_1}{\partial y_N} \\ \vdots & \vdots & & \vdots \\ \dfrac{\partial\phi_N}{\partial t} & \dfrac{\partial\phi_N}{\partial y_2} & \cdots & \dfrac{\partial\phi_N}{\partial y_N} \end{bmatrix} \quad (\text{using } y_1 = t)$$

$$= \begin{bmatrix} a_1 & * & \cdots & * \\ a_2 & & & \\ \vdots & & I_{N-1} & \\ a_N & & & \end{bmatrix}, \text{ at } y = x^0, (\text{by (A.28)}).$$

Thus,

$$\det D_y\phi(0, y_2, \ldots, y_N, y_1)|_{y=x^0} = a_1 \neq 0,$$

and the map (A.26) is invertible in a small open set around $y = x^0$. So (A.26) is a well-defined local coordinate transformation.

Since the variables $y_2, y_3, \cdots, y_N$ in $\phi(0, y_2, \ldots, y_N, y_1)$ are just initial conditions (such that $x_2(0) = x_2^0 = y_2, x_3(0) = x_3^0 = y_3, \ldots, x_N(0) = x_N^0 = y_N$), we have

$$\frac{dy_2}{dt} = 0, \frac{dy_3}{dt} = 0, \ldots, \frac{dy_N}{dt} = 0. \tag{A.29}$$

Combining (A.27) with (A.29), we have completed the proof.   □

Corollary A.19   Let x^0 be a nonequilibrium point of the differential equation $\dot{x} = f(x)$. Then in a neighborhood of x^0, there exist $N - 1$ independent integrals of motion, i.e., $N - 1$ functions $F_1(x), F_2(x), \ldots, F_{n-1}(x)$ such that

$$\frac{d}{dt} F_j(x(t)) = 0, \qquad j = 1, 2, \ldots, N - 1,$$

and

$$\sum_{j=1}^{N-1} \lambda_j D_x F_j(x) \not\equiv 0, \text{ for } x \text{ sufficiently close to } x^0,$$

for any $\lambda_1, \lambda_2, \ldots, \lambda_{N-1} \in \mathbb{R}$.

Proof. The coordinate functions $y_2, y_3, \ldots, y_N$ in the flow box theorem are constant in a neighborhood of x^0. Thus, they are integrals of motion.   □

Corollary A.20   Let x^0 be a nonequilibrium point of the differential equation $\dot{x} = f(x)$. Assume that $F(x)$ is an integral of motion such that F is nondegenerate near x^0, i.e., $D_x F(x)|_{x=x^0} \neq 0$. Then there exists a local coordinate system such that the differential equation $\dot{x} = f(x)$ is transformed to

$$\dot{y}_1 = 1; \; y_2 = F(x) \text{ and } \dot{y}_2 = 0; \; \dot{y}_3 = \dot{y}_4 = \cdots = \dot{y}_N = 0. \tag{A.30}$$

Proof. We know that y_1 is the same as the variable t, but $F(x)$ does not change along a trajectory. So $F(x)$ is independent of the y_1 variable. We have

$$D_y F(y) = \left[\frac{\partial F}{\partial y_1}, \frac{\partial F}{\partial y_2}, \ldots, \frac{\partial F}{\partial y_N} \right] \neq [0, 0, \ldots, 0],$$

by the nondegeneracy of F. Since $\frac{\partial F}{\partial y_1} = 0$, one of the $\frac{\partial F}{\partial y_2}, \ldots, \frac{\partial F}{\partial y_N}$ must be nonzero. Transform the coordinate system such that $\frac{\partial F}{\partial y_2} \neq 0$. Therefore, $y_2 = F(x)$ is uniquely solvable locally.   □

This local coordinate system y is called the flow box coordinates.

We now consider periodic solutions of an autonomous differential equation. The solution $x(t)$ of

$$\begin{cases} \dot{x}(t) = f(x(t)), & t > 0, \\ x(0) = \xi \in \mathbb{R}^N, \end{cases} \tag{A.31}$$

is denoted as $\phi(t, \xi)$. If this solution is periodic, then there exists a $T > 0$ such that

$$\phi(t + T, \xi) = \phi(t, \xi), \text{ for all } t > 0. \tag{A.32}$$

There is a smallest $T > 0$ satisfying (A.30) if $\boldsymbol{\phi}(t, \boldsymbol{\xi})$ is not a constant state. We call the smallest such $T > 0$ the *period* of the solution $\boldsymbol{\phi}(t, \boldsymbol{\xi})$.

Lemma A.21   ([54, p. 130]) The solution $\boldsymbol{\phi}(t, \boldsymbol{\xi})$ is a periodic solution with period T if and only if

$$\boldsymbol{\phi}(T, \boldsymbol{\xi}) = \boldsymbol{\xi}. \qquad (A.33)$$

Proof of Sufficiency. If (A.33) is satisfied, then

$$\begin{aligned} \boldsymbol{\phi}(t + T, \boldsymbol{\xi}) &= \boldsymbol{\phi}(t, \boldsymbol{\phi}(T, \boldsymbol{\xi})) \qquad \text{(semigroup property)} \\ &= \boldsymbol{\phi}(t, \boldsymbol{\xi}), \ \text{ for any } t > 0, \end{aligned}$$

so (A.32) is satisfied and the solution is periodic. (Necessity) If $\boldsymbol{\phi}(t, \boldsymbol{\xi})$ satisfies (A.32), then set $t = 0$ therein, we obtain

$$\boldsymbol{\phi}(T, \boldsymbol{\xi}) = \boldsymbol{\phi}(0, \boldsymbol{\xi}) = \boldsymbol{\xi},$$

so (A.33) is satisfied. $\qquad\qquad\square$

Definition A.22   Let $\boldsymbol{\phi}(t, \boldsymbol{\xi})$ be a periodic solution of (A.31) with period T. The matrix $D_{\boldsymbol{x}}\boldsymbol{\phi}(t, \boldsymbol{x})|_{\boldsymbol{x}=\boldsymbol{\xi}}$ is called the *monodromy* matrix at $\boldsymbol{\xi}$, and its eigenvalues are called the *characteristic multipliers* of the periodic solution $\boldsymbol{\phi}(t, \boldsymbol{\xi})$. $\qquad\qquad\square$

Lemma A.23   ([54, p. 130]) The monodromy matrix $D_{\boldsymbol{x}}\boldsymbol{\phi}(t, \boldsymbol{\xi})$ has 1 as its characteristic multiplier with eigenvector $\boldsymbol{f}(\boldsymbol{\xi})$.

Proof. Differentiating the semigroup relation

$$\frac{d}{dt}\boldsymbol{\phi}(\tau, \boldsymbol{\phi}(t, \boldsymbol{x})) = \frac{d}{dt}\phi(t + \tau, \boldsymbol{x})$$

yields

$$D_{\boldsymbol{y}}\boldsymbol{\phi}(\tau, \boldsymbol{\phi}(\tau, \boldsymbol{x})) = \frac{d}{dt}\boldsymbol{\phi}(t, \boldsymbol{x}) = \frac{d}{dt}\boldsymbol{\phi}(t + \tau, \boldsymbol{x}) \qquad (A.34)$$

Set $t = 0$, $\tau = T$ and $\boldsymbol{x} = \boldsymbol{\xi}$ in (A.34). We obtain

$$D_{\boldsymbol{y}}\boldsymbol{\phi}(T, \boldsymbol{\xi})\boldsymbol{f}(\boldsymbol{\xi}) = \boldsymbol{f}(\boldsymbol{\xi}). \qquad (A.35)$$

Since the periodic solution is not an equilibrium point, $\boldsymbol{f}(\boldsymbol{\xi}) \neq \boldsymbol{0}$. So $\boldsymbol{f}(\boldsymbol{\xi})$ is an eigenvector corresponding to the eigenvalue 1. $\qquad\qquad\square$

If $\xi_0 \in \mathbb{R}^n$ is an initial condition for a periodic solution $\phi(t, \xi_0)$, one is tempted to think that it might be possible to use the implicit function theorem (Theorem 1.7) to show that this T-periodic solution is unique in a neighborhood of ξ_0. Lemma A.23 shows that this is not true as $D_x\phi(T, \xi_0) - I_N$ is not invertible. In fact, (initial conditions for) periodic solutions are never isolated. The periodic solution itself constitutes a 1-dimensional manifold in a neighborhood of ξ_0 such that all points in the manifold correspond to an initial condition for a T-periodic solution.

In view of Lemma A.23 and the above, one can introduce instead a *cross section* to the periodic solution according to the following.

Definition A.24   Let $\phi(t, \xi_0)$ be a periodic solution of (A.31) with initial condition $\xi_0 \in \mathbb{R}^N$. Let $a \in \mathbb{R}^n$ be such that $a \cdot f(\xi_0) \neq 0$. Define the hyperplane passing ξ_0 with normal a as

$$\sum = \{x \in \mathbb{R}^n \mid a \cdot (x - \xi_0) = 0\}.$$

For any $\xi \in U \subseteq \sum$, for a small neighborhood U of ξ_0, let $\phi(\mathcal{T}(\xi), \xi) \in \sum$ be the point on $\phi(t, \xi)$ with the smallest $\mathcal{T}(\xi) > 0$. The map

$$P: \xi \in U \longrightarrow \phi(\mathcal{T}(\xi), \xi) \in \sum$$

is called the *Poincaré map* and $\mathcal{T}(\xi)$ is called the *first return time*. The hyperplane $\sum$ is called the *Poincaré section*.   □

One can see that Definition A.24 is based on very *geometric* ideas.

One can show that the Poincaré map is *smooth* ([54, Lemma V.2.4, pp. 133–132]). We present a few more properties of the Poincaré map in the following.

Lemma A.25   ([54, Lemma V.2.5, p. 132]) Let the characteristic multipliers of the monodromy matrix $D_x\phi(t, \xi_0)$ be 1, $\lambda_2, \lambda_3, \ldots, \lambda_N$ for a periodic solution $\phi(t, \xi_0)$. Then the eigenvalues of the linearized Poincaré map $D_y P$ at ξ_0 are $\lambda_2, \lambda_3, \ldots, \lambda_N$.

Proof. Make a coordinate transformation such that $\xi_0 = 0$ and $f(\xi_0) = (1, 0, 0, \ldots, 0)^T$. Thus, $\sum$ corresponds to the hyperplane $x_1 = 0$. By Lemma A.26, the monodormy matrix $M \equiv D_x\phi(T, \xi_0)$ has an eigenvalue 1 with eigenvector $f(\xi_0) = (1, 0, 0, \ldots, 0)$. Thus, M must take the form

$$M = \begin{bmatrix} 1 & * & * & \cdots & * \\ 0 & & & & \\ 0 & & & M' & \\ \vdots & & & & \\ 0 & & & & \end{bmatrix}$$

in order to satisfy

$$M \begin{bmatrix} 1 \\ 0 \\ \vdots \\ 0 \end{bmatrix} = \begin{bmatrix} 1 \\ 0 \\ \vdots \\ 0 \end{bmatrix}.$$

Since M has eigenvalues $1, \lambda_2, \lambda_3, \ldots, \lambda_N$, the $(N-1) \times (N-1)$ matrix M' must have eigenvalues $\lambda_2, \lambda_3, \ldots, \lambda_N$, belonging to the linearized Poincaré map at (the projection on $\sum$ of the) point $\boldsymbol{\xi}_0$.
□

Lemma A.26 ([54, Lemma V.4.7., p. 134]) Let $F(\boldsymbol{\xi})$ be an integral as warranted by Corollary A.20. If F satisfies $D_x F(x)|_{x=\phi(t,\xi_0)} \neq \mathbf{0}$ for the periodic solution $\boldsymbol{\phi}(t, \boldsymbol{\xi}_0)$, then the monodromy matrix $M \equiv D_x\boldsymbol{\phi}(T, \boldsymbol{\xi}_0)$ has a left eigenvector $D_x F(\boldsymbol{\xi}_0)$ with eigenvalue 1. Consequently, M has eigenvalue 1 with multiplicity 2.

Proof. Since F remains constant on a trajectory,

$$F(\boldsymbol{\phi}(t, \boldsymbol{\xi})) = F(\boldsymbol{\xi}),$$

and so

$$F(\boldsymbol{\phi}(T, \boldsymbol{\xi})) = F(\boldsymbol{\xi}),$$
$$D_x F(\boldsymbol{\xi}_0) D_x\boldsymbol{\phi}(T, \boldsymbol{\xi}_0) = D_x F(\boldsymbol{\xi}_0);$$

this verifies that $D_x F(\boldsymbol{\xi}_0)$ is a left eigenvector of M. We now choose a coordinate system such that $\boldsymbol{f}(\boldsymbol{\xi}_0)$ is the column vector $(1, 0, \ldots, 0)^T$ (T: transpose) as in the proof of Lemma A.25 and $D_x F(\boldsymbol{\xi}_0)$ is the row vector $(0, 1, 0, \ldots, 0)$. (We note that the two vectors $\boldsymbol{f}(\boldsymbol{\xi}_0)$ and $D_x F(\boldsymbol{\xi}_0)$ are independent if both are viewed as column vectors.) Thus, M takes the form

$$M = \begin{bmatrix} 1 & * & * & * & \cdots & * \\ 0 & 1 & 0 & 0 & \cdots & 0 \\ 0 & * & * & * & \cdots & * \\ 0 & * & & & & \vdots \\ \vdots & \vdots & & & & \vdots \\ 0 & * & * & * & \cdots & * \end{bmatrix}.$$

The characteristic polynomial must satisfy

$$\det(M - \lambda I) = (\lambda - 1)^2 \cdot p(\lambda)$$

for some polynomial $p(\lambda)$ of degree $N-2$.

Therefore, $\lambda = 1$ is an eigenvalue of multiplicity 2.   □

We may now analyze the Poincaré map a little further using the geometric theory established above for a periodic solution with period T: $\boldsymbol{\phi}(t + T, \boldsymbol{\xi}_0) = \phi(t, \boldsymbol{\xi}_0)$ with an integral $F(\boldsymbol{x})$, the monodromy matrix $D_{\boldsymbol{x}}\boldsymbol{\phi}(T, \boldsymbol{\xi}_0)$ has eigenvalues $1, 1, \lambda_2, \lambda_3, \ldots, \lambda_N$. Let us now choose flow box coordinates $y_1, y_2, \ldots, y_N$ satisfying (A.30).

If $\sum$ is a Poincaré section of $\boldsymbol{\phi}(t, \boldsymbol{\xi}_0)$ passing $\boldsymbol{\xi}_0$, then the trajectory $\boldsymbol{\phi}(t, \boldsymbol{\xi}_0)$ lies on the integral surface $F(\boldsymbol{x}) = F(\boldsymbol{\xi}_0)$, as Fig. A.13 shows.

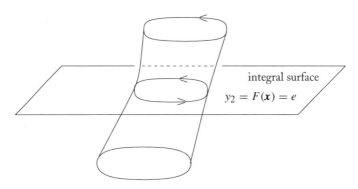

Figure A.13: A periodic orbit lying on an integral surface. This also illustrates the cylinder theorem (see Theorem A.27 to follow), where a family of periodic orbits exist. This figure is adapted from [54, Fig. V.E.4, p. 136].

Now, using the flow box coordinates $y_1, y_2, \cdots, y_N$ satisfying (A.30), i.e., $\dot{y}_1 = 1, y_2 = F(\boldsymbol{y})$, $\dot{y}_3 = \dot{y}_4 = \cdots = \dot{y}_N = 0$; cf. Crollary A.20. We can choose

$$\sum: \quad y_1 = 0, \quad \text{to be the Poincaré section.}$$

The geometry now look like what is displayed in Fig. A.14.

Then $F(\boldsymbol{x}) = F(\boldsymbol{\xi}_0) = F(\boldsymbol{\phi}(t, \boldsymbol{\xi}_0)) = y_2$. So y_2 is constant on the periodic trajectory $\boldsymbol{\phi}(t, \boldsymbol{\xi}_0)$. Call this constant e. Define

$$\sum_e \equiv \text{the intersection of } \sum \text{ with } y_2 = F(\boldsymbol{\xi}_0).$$

On $\sum_e$, define coordinates $\widehat{\boldsymbol{y}} = (y_3, y_4, \ldots, y_N)$ such that

$$\widehat{y}_1 = y_3, \quad \widehat{y}_2 = y_4, \ldots, \widehat{y}_{N-2} = y_N.$$

The Poincaré map thus becomes

$$P(e, \widehat{\boldsymbol{y}}) = (e, Q(e, \widehat{\boldsymbol{y}})), \qquad (e = F(\boldsymbol{\xi}_0) \text{ is fixed})$$

for some map $Q(e, \cdot)$ from a neighborhood N_e of the origin in $\sum_e$ to $\sum_e$, because $y_2 = F(\boldsymbol{\xi}_0) = e$.

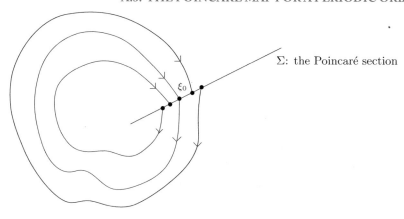

Figure A.14: Illustration of the Poincaré section.

Therefore, $Q(e, \hat{\mathbf{y}})$ is a further reduction of the Poincaré map from $\sum$ to $\sum_e$. Q is called the Poincaré map on the integral surface. Q satisfies

$$Q(e, 0) = 0$$

because $(e, 0)$ corresponds to $\boldsymbol{\xi}_0$ and is thus a fixed point of P.

When an integral of motion exists, then we often have a 1-parameter family of periodic orbits instead of a single periodic orbit, under some additional assumptions. This is the following famous theorem.

Theorem A.27   (The Cylinder Theorem) Let $\boldsymbol{\phi}(t, \boldsymbol{\xi}^0)$ denote a periodic orbit with period T of the differential equation $\dot{x} = f(x)$. Assume that there is a nondegenerate integral of motion $F(x)$. Assume further that the eigenvalues satisfy

$$\lambda_j \neq 1, \qquad j = 3, 4, \ldots, N;$$

for the monodromy matrix. Then there is a neighborhood of the periodic orbit $\boldsymbol{\phi}(t, \boldsymbol{\xi}^0)$, we have a 1-parameter family of periodic orbits parameterized by $e = F(x)$.

Proof. Any periodic orbit corresponds to a fixed point of the Poincaré map

$$P(e, \hat{\mathbf{y}}) = (e, Q(e, \hat{\mathbf{y}})).$$

The given periodic orbit satisfies

$$P(e_1, 0) = (e_1, 0).$$

We want to solve

$$P(e, \hat{\mathbf{y}}) = (e, Q(e, \hat{\mathbf{y}})) = (e, \hat{\mathbf{y}}), \text{ i.e., } Q(e, \hat{\mathbf{y}}) = \hat{\mathbf{y}}.$$

Applying the implicit function theorem to

$$g(e, z) = Q(e, z) - z, \text{ where } z \in \mathbb{R}^{N-2},$$

we have, near $e = e_1$,

$$D_z g(z)|_{z=0} = D_z Q(e, z) - D_z z = D_z Q(e, z) - I_{N-2}.$$

When $e = e_1$, $D_z Q(e_1, z)$ has eigenvalues $\lambda_2, \lambda_3, \ldots, \lambda_N$, none of which is equal to 1. So, the $(N - 2) \times (N - 2)$ matrix on the right-hand side of (6) is invertible when $e = e_1$. Therefore, $g(e, z) = 0$ has a unique solution z for each given e near $e = e_1$. Such a z is a fixed point of $Q(e, \cdot)$ and thus corresponds to a periodic orbit. $\qquad\square$

See an illustration in Fig. A.13.

The dynamic properties of the continuous-time autonomous system can be studied in terms of the discrete Poincaré map P. For nonautonomous system (A.1) with periodic forcing and other cases, it is also possible to define certain discrete Poincaré maps. The chaotic behaviors of such continuous-time systems are then analyzed through P using Melnikov's method [53], Smale's horseshoe, and other ideas. We refer the readers to Guckenheimer and Holmes [30], Meyer and Hall [54] and Wiggins [69] for some further study.

APPENDIX B

Chaotic Vibration of the Wave Equation due to Energy Pumping and van der Pol Boundary Conditions

The onset of chaotic phenomena in systems governed by nonlinear partial differential equations (PDEs) has fascinated scientists and mathematicians for many centuries. The most famous case in point is the Navier–Stokes equations and related models in fluid dynamics, where the occurrence of turbulence in fluids is well accepted as a chaotic phenomenon. Yet despite the diligence of numerous, most brilliant minds of mankind, and the huge amount of new knowledge gained through the vastly improved computational and experimental methods and facilities, at present, we still have not been able to rigorously prove that turbulence is indeed chaotic in a certain universal mathematical sense.

Nevertheless, for certain special partial differential equations, it is possible to rigorously prove the occurrence of chaos under certain given conditions. Here we mention the model of a vibrating string with nonlinear boundary conditions as studied by G. Chen, S.-B. Hsu, Y. Huang, J. Zhou, etc., in [10, 11, 12, 13, 14, 15, 16, 18, 36, 40].

B.1 THE MATHEMATICAL MODEL AND MOTIVATIONS

A linear wave equation

$$\frac{1}{c^2}\frac{\partial^2}{\partial t^2}w(x,t) - \frac{\partial^2}{\partial x^2}w(x,t) = 0, \qquad 0 < x < L, t > 0, \tag{B.1}$$

describes the propagation of waves on an interval of length L. For convenience, set the wave speed $c = 1$ and the length $L = 1$ in (B.1) as such values have no essential effect as far as the mathematical analysis of chaotic vibration here is concerned. We thus have

$$w_{tt}(x,t) - w_{xx}(x,t) = 0, \qquad 0 < x < 1, \quad t > 0. \tag{B.2}$$

The two initial conditions are

$$w(x,0) = w_0(x), \quad w_t(x,0) = w_1(x), \qquad 0 < x < 1. \tag{B.3}$$

At the right-end $x = 1$, assume a nonlinear boundary condition

$$w_x(1, t) = \alpha w_t(1, t) - \beta w_t^3(1, t); \qquad t > 0, \quad \alpha, \beta > 0. \tag{B.4}$$

At the left-end $x = 0$, we choose the boundary condition to be

$$w_t(0, t) = -\eta w_x(0, t), \qquad t > 0; \quad \eta > 0, \quad \eta \neq 1. \tag{B.5}$$

Remark B.1   Equation (B.5) says that negative force is fedback to the velocity at $x = 0$. An alternate choice would be

$$w_x(0, t) = -\eta w_t(0, t), \qquad t > 0; \quad \eta > 0, \quad \eta \neq 1,$$

which says negative velocity is fedback to force.   □

Remark B.2   An ordinary differential equation, called the van der Pol oscillator, is important in the design of classical servomechanisms:

$$\ddot{x} - (\alpha - \beta \dot{x}^2)\dot{x} + kx = 0; \qquad \alpha, \beta > 0, \tag{B.6}$$

where $x = x(t)$ is proportional to the electric current at time t on a circuit equipped with a *van der Pol device*. Then the energy at time t is $E(t) = \frac{1}{2}(\dot{x}^2 + kx^2)$ and

$$\frac{d}{dt}E(t) = \dot{x}(\ddot{x} + kx) = \dot{x}^2(\alpha - \beta \dot{x}^2),$$

so we have

$$E'(t) \begin{cases} \geq 0 & \text{if } |\dot{x}| \leq (\alpha/\beta)^{1/2}, \\ < 0 & \text{if } |\dot{x}| > (\alpha/\beta)^{1/2} \end{cases} \tag{B.7}$$

which is a desired *self-regulation effect*, i.e., energy will increase when $|\dot{x}|$ is small (which is unfit for operations), and energy will decrease when $|\dot{x}|$ is large in order to prevent electric current surge which may destroy the circuit. (This self-regulating effect is also called *self-excitation*.) A second version of the van der Pol equation is

$$\ddot{x} - (\alpha - 3\beta x^2)\dot{x} + kx = 0, \tag{B.8}$$

which may be regarded as a differentiated version of (B.6), satisfying a regulation effect similar to (B.7). Neither (B.6) nor (B.8) has any chaotic behavior as *the solutions tend to limit cycles according to the Poincaré–Bendixon Theorem.* However, when a *forcing term* $A \cos(\omega t)$ is added to the right-hand side of (B.6) or (B.8), solutions display chaotic behavior when the parameters A and ω enter a certain regime [30, 48].   □

With (B.4) and (B.5), we have, by (B.2), (B.4) and (B.5),

$$
\begin{aligned}
\frac{d}{dt} E(t) &= \frac{d}{dt} \int_0^1 \left[\frac{1}{2} w_x^2(x, t) + w_t^2(x, t) \right] dt \\
&= \int_0^1 [w_x(x, t) w_{xt}(x, t) + w_t(x, t) w_{tt}(x, t)] dx \\
&= \int_0^1 [w_x(x, t) w_{xt}(x, t) + w_t(x, t) w_{xx}(x, t)] dx \\
&\quad (\Rightarrow \text{ integration by parts}) \\
&= w_t(x, t) w_x(x, t)|_{x=0}^{x=1} \\
&= \eta w_x^2(0, t) + w_t^2(1, t)[\alpha - \beta w_t^2(1, t)].
\end{aligned}
\tag{B.9}
$$

The contribution $\eta w_x^2(0, t)$ above, due to (B.5), is always nonnegative. Thus, we see that the effect of (B.5) is to cause energy to increase. For this reason, the boundary condition (B.5) is said to be *energy-injecting* or *energy-pumping*. On the other hand, we have

$$
w_t^2(1, t)[\alpha - \beta w_t^2(1, t)] \begin{cases} \geq 0 & \text{if } |w_t(1, t)| \leq (\alpha/\beta)^{1/2}, \\ < 0 & \text{if } |w_t(1, t)| > (\alpha/\beta)^{1/2}, \end{cases}
\tag{B.10}
$$

so the contribution of the boundary condition (B.4) to (B.9) is *self-regulating* because (B.10) works in exactly the same way as (B.7). Thus, we call (B.4) a *van der Pol, self-regulating*, or *self-excitating*, boundary condition. Intuitively speaking, with the boundary condition (B.5) alone (and with the right-end boundary condition (B.4), replaced by a conservative boundary condition such as $w(1, t) = 0$ or $w_x(1, t) = 0$ for all $t > 0$), it causes the well-known *classical linear instability*, namely, the energy grows with an exponential rate:

$$
E(t) = \mathcal{O}(e^{kt}), \quad k = \frac{1}{2} \ln \left(\left| \frac{1 + \eta}{1 - \eta} \right| \right) > 0.
\tag{B.11}
$$

However, the self-regulating boundary condition (B.4) can hold the instability (B.11), *partly* in check by its regulation effect, for a large class of *bounded initial states* with bounds depending on the parameters α, β and η. When α, β and η match in a certain regime, chaos happens, which could be viewed as a *reconciliation between linear instability and nonlinear self-regulation*. Overall, there is a richness of nonlinear phenomena, including the following: the existence of *asymptotically periodic solutions, hysteresis, instability* of the type of unbounded growth, and *fractal invariant sets*.

A basic approach for the problems under consideration in this section is the *method of characteristics*. Let u and v be the *Riemann invariants* of (B.2) defined by

$$
\begin{aligned}
u(x, t) &= \frac{1}{2}[w_x(x, t) + w_t(x, t)], \\
v(x, t) &= \frac{1}{2}[w_x(x, t) - w_t(x, t)].
\end{aligned}
\tag{B.12}
$$

Then u and v satisfy a diagonalized first order linear hyperbolic system

$$\frac{\partial}{\partial t}\begin{bmatrix} u(x,t) \\ v(x,t) \end{bmatrix} = \begin{bmatrix} 1 & 0 \\ 0 & -1 \end{bmatrix}\frac{\partial}{\partial x}\begin{bmatrix} u(x,t) \\ v(x,t) \end{bmatrix}, \qquad 0 < x < 1, \quad t > 0, \tag{B.13}$$

with initial conditions

$$\left.\begin{aligned} u(x,0) = u_0(x) &\equiv \frac{1}{2}[w_0'(x) + w_1(x)], \\ v(x,0) = v_0(x) &\equiv \frac{1}{2}[w_0'(x) - w_1(x)], \end{aligned} \quad 0 < x < 1. \right\} \tag{B.14}$$

The boundary condition (B.4), after converting to u and v and simplifying, becomes

$$u(1,t) = F_{\alpha,\beta}(v(1,t)), \qquad t > 0, \tag{B.15}$$

where the relation $u = F_{\alpha,\beta}(v)$ is defined implicitly by

$$\beta(u - v)^3 + (1 - \alpha)(u - v) + 2v = 0; \qquad \alpha, \beta > 0. \tag{B.16}$$

Remark B.3   For (B.16), we know that

(i) when $0 < \alpha \leq 1$, for each $v \in \mathbb{R}$, there exists a unique $u \in \mathbb{R}$;

(ii) when $\alpha > 1$, for each $v \in \mathbb{R}$, in general there may exist two or three distinct $u \in \mathbb{R}$ satisfying
     (B.16). Thus, $u = F_{\alpha,\beta}(v)$ is not a function relation.

Only case (i) will be treated in Section B.2 while for case (ii), which contains hysteresis, the interested
reader may refer to [14].   ☐

The boundary conditions (B.2), by (B.12), becomes

$$v(0,t) = G_\eta(u(0,t)) \equiv \frac{1 + \eta}{1 - \eta}u(0,t), \qquad t > 0. \tag{B.17}$$

Equations (B.16) and (B.17) are, respectively, the *wave-reflection relations* at the right-end $x = 1$
and the left-end $x = 0$. The reflection of characteristics is depicted in Fig. B.1.

Assume that $F_{\alpha,\beta}$ is well defined. Then a solution (u, v) of the system (B.13), (B.14), (B.15)
and (B.17) can be expressed as follows:

For $0 \leq x \leq 1$ and $t = 2k + \tau$, with $k = 0, 1, 2, \ldots$, and $0 \leq \tau < 2$,

$$u(x,t) = \begin{cases} (F \circ G)^k(u_0(x + \tau)), & \tau \leq 1 - x, \\ G^{-1} \circ (G \circ F)^{k+1}(v_0(2 - x - \tau)), & 1 - x < \tau \leq 2 - x, \\ (F \circ G)^{k+1}(u_0(\tau + x - 2)), & 2 - x < \tau \leq 2; \end{cases}$$

$$v(x,t) = \begin{cases} (G \circ F)^k(v_0(x - \tau)), & \tau \leq x, \\ G \circ (F \circ G)^k(u_0(\tau - x)), & x < \tau \leq 1 + x, \\ (G \circ F)^{k+1}(v_0(2 + x - \tau)), & 1 + x < \tau \leq 2, \end{cases} \tag{B.18}$$

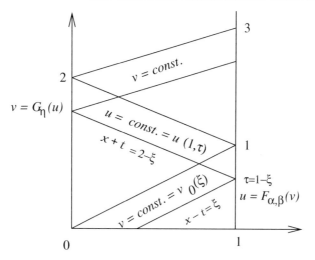

Figure B.1: Reflection of characteristics.

where in the above, $F = F_{\alpha,\beta}$ and $G = G_\eta$, and $(G \circ F)^k$ represents the k-th iterate of the map $G \circ F$. From now on, we often abbreviate $F_{\alpha,\beta}$ and G_η, respectively, as F and G, in case no ambiguities will occur. We call the map $G_\eta \circ F_{\alpha,\beta}$, naturally, the *composite reflection relation*. This map $G_\eta \circ F_{\alpha,\beta}$ can be regarded as the *Poincaré section* of the PDE system because we can essentially construct the solution from $G_\eta \circ F_{\alpha,\beta}$ using (B.18).

From (B.18), it becomes quite apparent that the solutions $(u(x, t), v(x, t))$ will manifest chaotic behavior when the map $G \circ F$ is chaotic, in the sense of Devaney cf. Definition 6.27 in Chapter 6 and [20, p. 50], for example. We proceed with the discussion in the following section. The main source of reference is [13].

B.2   CHAOTIC VIBRATION OF THE WAVE EQUATION

As mentioned in Remark B.1, when $0 < \alpha \leq 1$, for each $v \in \mathbb{R}$ there exists a unique $u \in \mathbb{R}$ such that $u = F_{\alpha,\beta}(v)$. Therefore, the solution (u, v) to (B.13), (B.14), (B.15) and (B.17) is unique. When the initial condition (u_0, v_0) is sufficiently smooth satisfying compatibility conditions with the boundary conditions, then (u, v) will also be C^1-smooth on the spatiotemporal domain.

Let α and β be fixed, and let $\eta > 0$ be the only parameter that varies. To aid understanding, we include a sample graph of the map $G_\eta \circ F_{\alpha,\beta}$, with $\alpha = 1/2$, $\beta = 1$, and $\eta = 0.552$, in Fig. B.2. We only need to establish that $G_\eta \circ F_{\alpha,\beta}$ is chaotic, because $F_{\alpha,\beta} \circ G_\eta$ is topologically conjugate to $G_\eta \circ F_{\alpha,\beta}$ through

$$F_{\alpha,\beta} \circ G_\eta = G_\eta^{-1} \circ (G_\eta \circ F_{\alpha,\beta}) \circ G_\eta$$

and, thus, the iterates $(F \circ G)^k$ or $(F \circ G)^{k+1}$ appearing in (B.18) do not need to be treated separately.

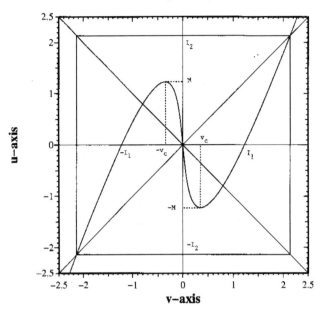

Figure B.2: The graph of $u = G_\eta \circ F_{\alpha,\beta}(v)$. (Here $\alpha = 0.5$, $\beta = 1$, $\eta = 0.552$ are used.) Note that
$\pm I_1 = v$-axis nonzero intercepts, $I_1 = [(1 + \alpha/\beta]^{1/2}$;
$\pm v_c = $ (local) critical points, $v_c = [(2 - \alpha)/3][(1 + \alpha)/(3\beta)]^{1/2}$;
$\pm M = $ local extremum values, $M = \frac{1+\eta}{1-\eta}\frac{1+\alpha}{3}[(1 + \alpha/3\beta]^{1/2}$;
$\pm I_2 = v$-values where the curve intersects with the line $u - v = 0$,
and $[-I_2, I_2] \times [-I_2, I_2]$ is an invariant square when $M \leq I_2$.

We note the following bifurcations: For fixed α: $0 < \alpha \leq 1$ and $\beta > 0$, let $\eta \in (0, 1)$ be varying.

(1) *Period-doubling bifurcation* (Theorem 4.1).
    Define

$$h(v, \eta) = -G_\eta \circ F(v)$$

and let

$$v_0(\eta) \equiv \eta[(1 + \eta)/2][(\alpha + \eta)/\beta]^{1/2}$$

which, for each η, represents a fixed point of h, i.e.,

$$h(v_0(\eta), \eta) = v_0(\eta).$$

Then the algebraic equation

$$\frac{1}{2}\left(\frac{1+\alpha\eta}{3\beta\eta}\right)^{1/2}\left[\frac{1+(3-2\alpha)\eta}{3\eta}\right] = \frac{1+\eta}{2}\left(\frac{\alpha+\eta}{\beta}\right)^{1/2} \qquad (B.19)$$

has a unique solution $\eta = \eta_0 : \; 0 < \eta_0 \leq \boldsymbol{\eta}_H$, where

$$\boldsymbol{\eta}_H \equiv \left(1 - \frac{1+\alpha}{3\sqrt{3}}\right) \bigg/ \left(1 + \frac{1+\alpha}{3\sqrt{3}}\right) \tag{B.20}$$

satisfying

$$\frac{\partial}{\partial v} h_1(v, \eta) \bigg|_{\substack{v=v_0(\eta_0) \\ \eta=\eta_0}} = -1 \tag{B.21}$$

which is the primary necessary condition for period-doubling bifurcation to happen, at $v = v_0(\eta_0)$, $\eta = \eta_0$. Furthermore, the other "accessory" conditions are also satisfied, and the bifurcationed period-2 solutions are attracting.

Consequently, there is a period-doubling route to chaos, as illustrated in the orbit diagram in Fig. B.3.

(2) *Homoclinic orbits* (cf. Chapter 5).
Let $\boldsymbol{\eta}_H$ be given by (B.19). If

$$\boldsymbol{\eta}_H \leq \eta < 1, \tag{B.22}$$

then $M \geq I_1$ (cf. Fig. B.2) and, consequently, the repelling fixed point 0 of $G_\eta \circ F$ has homoclinic orbits. Furthermore, if $\eta = \boldsymbol{\eta}_H$, then there are *degenerate homoclinic orbits* (and, thus, homoclinic bifurcations [20, p. 125]).

When $M > I_2$; cf. Fig. B.2, then $[-I_2, I_2] \times [-I_2, I_2]$ is no longer an invariant square for the map $G \circ F$. What happens is exactly similar to the case of the quadratic map $f_\mu(x) = \mu x(1 - x)$, for $0 \leq x \leq 1$, when $\mu > 4$ because part of the graph of f_μ will protrude above the unit square. See Fig. B.4. It is easy to see that now the map $G \circ F$ has a Cantor-like fractal invariant set Λ on the interval $[-I_2, I_2]$, where $\Lambda = \bigcap_{j=1}^{\infty} (G \circ F)^k([-I_2, I_2])$. All the other points outside Λ are eventually mapped to $\pm\infty$ as the number of iterations increases.
We furnish a PDE example below.

Example B.4 ([13, p. 435, Example 3.3]) Consider (B.13), (B.14), (B.15) and (B.17), where we choose

$$\alpha = 0.5, \beta = 1, \eta = 0.525 \approx \boldsymbol{\eta}_H, \text{ satisfying (B.22)},$$
$$w_0(x) = 0.2 \sin\left(\frac{\pi}{2}x\right), \quad w_1(x) = 0.2 \sin(\pi x), \quad x \in [0, 1].$$

Two spatiotemporal profiles of u and v are plotted, respectively, in Figs. B.5 and B.6. Their rugged outlooks manifest chaotic vibration. ☐

Miscellaneous remarks

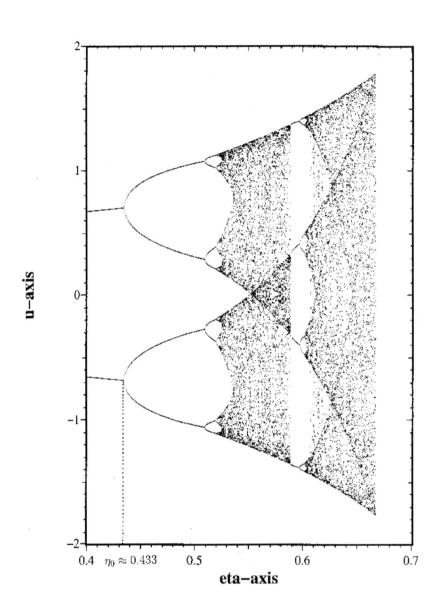

Figure B.3: The orbit diagram of $G_\eta \circ F_{\alpha,\beta}$, where $\alpha = 0.5$, $\beta = 1$, and η varies in [0.4, 2/3]. Note that the first period-doubling occurs near $\eta_0 \approx 0.433$, agreeing with the computational result of the solution η_0 satisfying equation (B.21). (Reprinted from [13, p. 433, Fig. 3], courtesy of World Scientific, Singapore.)

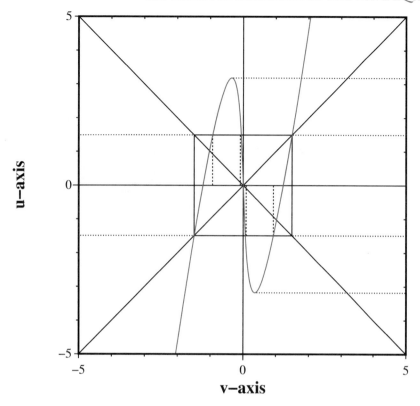

Figure B.4: The graph of $G_\eta \circ F_{\alpha,\beta}$ with $\alpha = 0.5$, $\beta = 1$ and $\eta = 0.8$. Note that here $M > I_2$ (cf. Fig. B.2) and $[-I_2, I_2] \times [-I_2, I_2]$ is no longer an invariant square for $G_\eta \circ F_{\alpha,\beta}$. On $[-I_2, I_2]$, what $G_\eta \circ F_{\alpha,\beta}$ has is a Cantor-like fractal invariant set.

(1) In this subsection, we have illustrated only the case $0 < \eta < 1$. When $\eta > 1$, the results are similar. See [13].

(2) With the nonlinear boundary condition (B.4), we can only establish that u and v are chaotic. From this, we can then show that w_x and w_t, i.e., the gradient of w, are also chaotic by a natural topological conjugacy, see [12, Section 5]. However, w itself is not chaotic because w is the time integral of w_t, which smooths out the oscillatory behavior of w_t.
In order to have chaotic vibration of w, one must use a differentiated boundary condition; see [13, Section 6]. This is actually an analog of (B.8).

(3) When the initial data (u_0, v_0) takes values outside the invariant square $[-I_2, I_2] \times [I_2, I_2]$, then part of u and v will diverge to $\pm\infty$ as $t \to \infty$. This behavior belongs to classical unbounded instability.

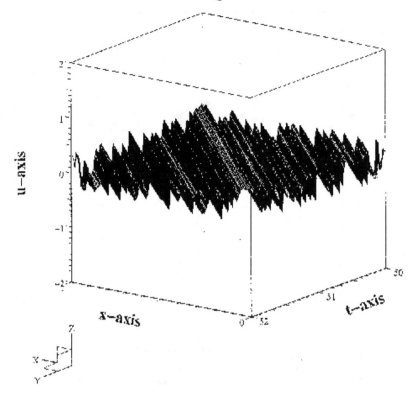

Figure B.5: The spatiotemporal profile of the u-component for Example 1.1, for $t \in [50, 52], x \in [0, 1]$. (Reprinted from [13, p. 435, Fig. 7], courtesy of World Scientific, Singapore.)

Further studies of the chaotic vibration of the wave equation for the case when there is boundary hysteresis or the case of nonisotropic-spatio temporal chaos and others may be found in [10, 11, 12, 15] and [16].

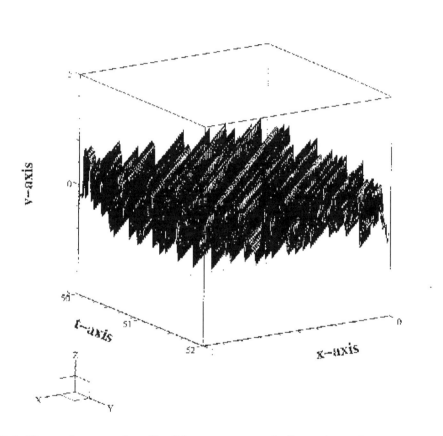

Figure B.6: The spatiotemporal profile of the v-component for Example 1.1, for $t \in [50, 52]$, $x \in [0, 1]$. (Reprinted from [13, p. 435, Fig. 8], courtesy of World Scientific, Singapore.)

Bibliography

[1] R.H. Abraham and C.D. Shaw, *Dynamics–The Geometry of Behavior, Part 4: Bifurcation Behavior*, Aerial Press, Inc., Santa Cruz, CA, 1988. Cited on page(s) 55

[2] V.S. Afraimovich and S.-B. Hsu, *Lectures on chaotic dynamical systems* (AMS/IP Studies in Advanced Mathematics), American Mathematical Society, Providence, R.I., 2002. Cited on page(s) xiii, 61

[3] J. Banks, J. Brooks, G. Cairns and P. Stacey, On Devaney's definition of chaos, *Amer. Math. Monthly* **99** (1992), 332–334. DOI: 10.2307/2324899 Cited on page(s) 75

[4] J. Benhabib and R.H. Day, Rational choice and erratic behaviour, *Rev. Econom. Stud.* **48** (1981), 459–471. DOI: 10.2307/2297158 Cited on page(s) 152

[5] J. Benhabib and R.H. Day, A characterization of erratic dynamics in the overlapping generations model, *J. Econom. Dynamics Control* **4** (1982), 37–55. DOI: 10.1016/0165-1889(82)90002-1 Cited on page(s) 152

[6] L. Block, Homoclinic points of mappings of the interval, *Proc. Amer. Math. Soc.* **72** (1978), 576–580. DOI: 10.1090/S0002-9939-1978-0509258-X Cited on page(s)

[7] L. Block and W.A. Coppel, *Dynamics in One Dimension*, Lecture Notes in Mathematics, Vol. 1513, Springer Verlag, New York-Heidelberg-Berlin, 1992. Cited on page(s) 20

[8] L. Block, J. Guckenheimer, M. Misiurewicz, and L.-S. Young, Periodic points and topological entropy of one dimensional maps, in *Lecture Notes in Mathematica*, Vol. 819, Springer-Verlag, New York-Heidelberg-Berlin, 1980, 18–34. Cited on page(s) 33

[9] U. Burkart, Interval mapping graphs and periodic points of continuous functions, *J. Combin. Theory Ser. B* **32** (1982), 57–68. DOI: 10.1016/0095-8956(82)90076-4 Cited on page(s) 33

[10] G. Chen, S.B. Hsu and T.W. Huang, Analyzing the displacement terms memory effect to prove the chaotic vibration of the wave equation, *Int. J. Bifur. Chaos* **12** (2002), 965–981. DOI: 10.1142/S0218127402005741 Cited on page(s) 205, 214

[11] G. Chen, S.B. Hsu and J. Zhou, Linear superposition of chaotic and orderly vibrations on two serially connected strings with a van der Pol joint, *Int. J Bifur. Chaos* **6** (1996), 1509–1527 DOI: 10.1142/S021812749600076X Cited on page(s) 205, 214

[12] G. Chen, S.B. Hsu and J. Zhou, Chaotic vibrations of the one-dimensional wave equation due to a self-excitation boundary condition. Part I, controlled hysteresis, *Trans. Amer. Math. Soc.* **350** (1998), 4265–4311. DOI: 10.1090/S0002-9947-98-02022-4 Cited on page(s) 162, 205, 213, 214

[13] G. Chen, S.B. Hsu and J. Zhou, Chaotic vibrations of the one-dimensional wave equation due to a self-excitation boundary condition. Part II, energy injection, period doubling and homoclinic orbits, *Int. J. Bifur. Chaos* **8** (1998), 423–445. DOI: 10.1142/S0218127498001236 Cited on page(s) 205, 209, 211, 212, 213, 214, 215

[14] G. Chen, S.B. Hsu and J. Zhou, Chaotic vibrations of the one-dimensional wave equation due to a self-excitation boundary condition. Part III, natural hysteresis memory effects, *Int. J. Bifur. Chaos* **8** (1998), 447–470. DOI: 10.1142/S0218127498001236 Cited on page(s) 205, 208

[15] G. Chen, S.B. Hsu and J. Zhou, Snapback repellers as a cause of chaotic vibration of the wave equation with a van der Pol boundary condition and energy injection at the middle of the span, *J. Math. Phys.* **39** (1998), 6459–6489. DOI: 10.1063/1.532670 Cited on page(s) 109, 162, 205, 214

[16] G. Chen, S.B. Hsu and J. Zhou, Nonisotropic spatiotemporal chaotic vibration of the wave equation due to mixing energy transport and a van der Pol boundary condition, *Int. J. Bifur. Chaos* **12** (2002), 447–470. DOI: 10.1142/S0218127402005741 Cited on page(s) 205, 214

[17] G. Chen, T. Huang and Y. Huang, Chaotic behavior of interval maps and total variations of iterates, *Int. J. Bifur. Chaos* **14** (2004), 2161–2186. DOI: 10.1142/S0218127404010242 Cited on page(s) 28, 177

[18] G. Chen, T. Huang, J. Juang, and D. Ma, Unbounded growth of total variations of snapshots of the 1D linear wave equation due to the chaotic behavior of iterates of composite nonlinear boundary reflection relations, G. Chen et al. (ed.), in *Control of Nonlinear Distributed Parameter Systems*, Marcel Dekker Lectures Notes on Pure Appl. Math., New York, 2001, 15–43. Cited on page(s) 205

[19] K. Ciesielski and Z. Pogoda, On ordering the natural numbers or the Sharkovski theorem, *Amer. Math. Monthly* **115** (2008), no. 2, 159–165. Cited on page(s) 33

[20] R.L. Devaney, *An Introduction to Chaotic Dynamical Systems*, 2nd ed., Addison-Wesley, New York, 1989. Cited on page(s) xiii, 20, 36, 60, 75, 86, 209, 211

[21] P. Diamond, Chaotic behavior of systems of difference equations, *Int. J. Systems Sci.* **7** (1976), 953–956. DOI: 10.1080/00207727608941979 Cited on page(s) 153

[22] A. Dohtani, Occurrence of chaos in higher-dimensional discrete-time systems, *SIAM J. Appl. Math.* **52** (1992), 1707–1721 DOI: 10.1137/0152098 Cited on page(s) 152, 153

[23] B. Du, A simple proof of Sharkovsky's theorem, *Amer. Math. Monthly* **111** (2004), 595–599. DOI: 10.2307/4145161 Cited on page(s) 32, 33

[24] B. Du, A simple proof of Sharkovsky's theorem revisited, *Amer. Math. Monthly* **114** (2007), 152–155. Cited on page(s) 32, 33

[25] B. Du, A simple proof of Sharkovsky's theorem re-revisited, preprint. (Version 7, September 28, 2009.) Cited on page(s) 32

[26] J. Dugandji, *Topology*, Allyn and Bacon, Boston, Massachusetts, 1967. Cited on page(s) 79, 104

[27] J.-P. Eckmann, S.O. Kamphorst, D. Ruelle, and S. Ciliberto, Lyapunov exponent from time series, *Phys. Rev. A* **34** (1986), 4971–4979 http://mpej.unige.ch/~eckmann/ps_files/eckmannkamphorstruelle.pdf/. DOI: 10.1103/PhysRevA.34.4971 Cited on page(s) 68, 129

[28] K.J. Falconer, *The Geometry of Fractal Sets*, Cambridge University Press, 1985. Cited on page(s) 129, 136

[29] K.J. Falconer, *Fractal Geometry*, John Wiley and Sons, New York, 1990. Cited on page(s) 129, 135

[30] J. Guckenheimer and P. Holmes, *Nonlinear Oscillations, Dynamical Systems and Bifurcations of Vector Fields*, Springer, New York, 1983. Cited on page(s) xiii, 56, 204, 206

[31] J.K. Hale and H. Kocak, *Dynamics and Bifurcations*, Springer Verlag, New York-Heidelberg-Berlin, 1991. Cited on page(s) 56

[32] R. Hegger, H. Kantz and T. Schreiber, *Nonlinear Time Series Analysis*, TISEAN 3.0.1 (March 2007), http://en.wikipedia.org/wiki/Tisean/. Cited on page(s) 68

[33] C.W. Ho and C. Morris, A graph-theoretic proof of Sharkovsky's theorem on the periodic points of continuous functions, *Pacific J. Math.* **96** (1981), 361–370. Cited on page(s) 33

[34] Y. Huang, Growth rates of total variations of snapshots of the 1D linear wave equation with composite nonlinear boundary reflection, *Int. J. Bifur. Chaos* **13** (2003), 1183–1195. DOI: 10.1142/S0218127403007138 Cited on page(s)

[35] Y. Huang, A new characterization of nonisotropic chaotic vibrations of the one-dimensional linear wave equation with a van der Pol boundary condition, *J. Math. Anal. Appl.* **288** (2003), no. 1, 78–96. DOI: 10.1016/S0022-247X(03)00562-6 Cited on page(s)

[36] Y. Huang, Boundary feedback anticontrol of spatiotemporal chaos for 1D hyperbolic dynamical systems, *Int. J. Bifur. Chaos* **14** (2004), 1705–1723. DOI: 10.1142/S021812740401031X Cited on page(s) 205

[37] Y. Huang, G. Chen and D.W. Ma, Rapid fluctuations of chaotic maps on $\mathbb{R}^N$, *J. Math. Anal. Appl.* **323** (2006), 228–252. DOI: 10.1016/j.jmaa.2005.10.019 Cited on page(s) 160

[38] Y. Huang and Z. Feng, Infinite-dimensional dynamical systems induced by interval maps, *Dyn. Contin. Discrete Impuls. Syst. Ser. A Math. Anal.* **13** (2006), no. 3–4, 509–524. Cited on page(s) 177

[39] Y. Huang, X.M. Jiang and X. Zou, Dynamics in numerics: On a discrete predator-prey model, *Differential Equations Dynam. Systems* **16** (2008), 163–182. DOI: 10.1007/s12591-008-0010-6 Cited on page(s) 160

[40] Y. Huang, J. Luo and Z.L. Zhou, Rapid fluctuations of snapshots of one-dimensional linear wave equation with a van der Pol nonlinear boundary condition, *Int. J. Bifur. Chaos* **15** (2005), 567–580. DOI: 10.1142/S0218127405012223 Cited on page(s) 205

[41] Y. Huang and Y. Zhou, Rapid fluctuation for topological dynamical systems, *Front. Math. China* **4** (2009), 483–494. DOI: 10.1007/s11464-009-0030-8 Cited on page(s) 160

[42] W. Hurewicz and H. Wallman, *Dimension Theory*, revised edition, Princeton University Press, Princeton, New Jersey, 1996. Cited on page(s) 129

[43] G. Iooss and D.D. Joseph, *Elementary Stability and Bifurcation Theory*, Springer Verlag, New York-Heidelberg-Berlin, 1980. Cited on page(s) 55, 56

[44] C.G.J. Jacobi, Über die Figur des Gleichgewichts, *Poggendorff Annalen der Physik und Chemie* **32** (229), 1834. DOI: 10.1002/andp.18341090808 Cited on page(s) 55

[45] J. Kennedy and J.A. Yorke, Topological horseshoe, *Trans. Amer. Math. Soc.* **353** (2001), 2513–2530. DOI: 10.1090/S0002-9947-01-02586-7 Cited on page(s) 150

[46] Y.A. Kuznetsov, *Elements of Applied Bifurcation Theory* (Applied Mathematical Sciences), Springer, New York, 2010. Cited on page(s) 50

[47] P.-S. Laplace, *A Philosophical Essay on Probabilities*, translated from the 6th French edition by Frederick Wilson Truscott and Frederick Lincoln Emory, Dover Publications, New York, 1951. Cited on page(s) xi

[48] M. Levi, Qualitative analysis of the periodically forced relaxation oscillations, *Mem. Amer. Math. Soc.* **214** (1981), 1–147. DOI: 10.1007/BFb0086995 Cited on page(s) 206

[49] T.Y. Li and J.A. Yorke, Period three implies chaos, *Amer. Math. Monthly* **82** (1975), 985–992. DOI: 10.2307/2318254 Cited on page(s) xii, 87

[50] A. Lypunoff, Problemes general de la stabilitede mouvement, *Ann. Fac. Sci. Univ. Toulouse* **9** (1907), 203–475. Cited on page(s) 68

[51] F.R. Marotto, Snap-back repellers imply chaos in R^n, *J. Math. Anal. Appl.* **63** (1978), 199–223. DOI: 10.1016/0022-247X(78)90115-4 Cited on page(s) 106

[52] F.R. Marotto, On redefining a snap-back repeller, *Chaos, Solitons, and Fractals* **25** (2005), 25–28. DOI: 10.1016/j.chaos.2004.10.003 Cited on page(s) 106

[53] V.K. Melnikov, On the stability of the center for time periodic perturbations, *Trans. Moscow Math. Soc.* **12** (1963), 1–57. Cited on page(s) 204

[54] K.R. Meyer and G.R. Hall, *Introduction to Hamiltonian Dynamical Systems and the N-Body Problem*, Springer, New York, 1992. Cited on page(s) xiii, 195, 199, 200, 201, 202, 204

[55] M. Morse and G.A. Hedlund, Symbolic dynamics, *Amer. J. Math.* **60** (1938), 815–866. DOI: 10.2307/2371264 Cited on page(s) 109

[56] J. Moser, *Stable and Random Motions in Dynamical Systems*, Princeton University Press, Princeton, New Jersey, 1973. Cited on page(s) 124

[57] H. Poincaré, Sur l'équilibre d'une masse fluide animée d'un mouvement de rotation, *Acta Math.* **7** (1885), 259–380. DOI: 10.1007/BF02402204 Cited on page(s) 55

[58] C. Robinson, *Dynamical Systems, Stability, Symbolic Dynamics and Chaos*, CRC Press, Boca Raton, FL, 1995, pp. 67–69. Cited on page(s) xiii, 20, 30, 42, 56, 68

[59] R.C. Robinson, *An Introduction to Dynamical Systems: Continuous and Discrete*, Pearson Education Asia Limited and China Machine Press, 2005. Cited on page(s) 82

[60] A.M.   Rucklidge,   http://www.maths.leeds.ac.uk/~alastair/MATH3395/examples_2.pdf/. Cited on page(s) 68

[61] C.E. Shannon, A mathematical theory of communication, *Bell System Technical J.* **27** (1948), 379–423. DOI: 10.1145/584091.584093 Cited on page(s) 109

[62] A.N. Sharkovskii, Coexistence of cycles of a continuous mapping of a line into itself, *Ukrainian Math. J.* 1964. Cited on page(s) 32

[63] A.N. Sharkovsky, Y.L. Maistrenko and E.Y. Romanenko, *Difference Equations and Their Applications*, Ser. Mathematics and its Applications, **250**, Kluwer Academic Publisher, Dordrecht, 1993. Cited on page(s) 162, 164

[64] S. Smale, Diffeomorphisms with many periodic points, in *Differential and Combinatorial Topology*, S.S. Cairns (ed.), Princeton University Press, Princeton, New Jersey, 1963, pp. 63–80. Cited on page(s) 116, 124

[65] S. Smale, *The Mathematics of Time: Essays on Dynamical Systems, Economic Processes and Related Topics*, Springer, New York, 1980. Cited on page(s) 116

[66] P. Stefan, A theorem of Sarkovskii on the coexistence of periodic orbits of continuous endomorphisms of the real line, *Comm. Math. Phys.* **54** (1977), 237–248. DOI: 10.1007/BF01614086 Cited on page(s) 33

[67] Zhi-Ying Wen, *Mathematical Foundations of Fractal Geometry*, Advanced Series in Nonlinear Science, Shanghai Scientific and Technological Education Publishing House, Shanghai, 2000 (in Chinese). Cited on page(s) xiii, 129, 137

[68] S. Wiggins, *Global Bifurcations and Chaos: Analytical Methods*, Springer Verlag, New York-Heidelberg-Berlin, 1988. Cited on page(s) 56, 124

[69] S. Wiggins, *Introduction to Applied Nonlinear Dynamical Systems and Chaos*, 2nd ed., Springer, New York, 2003. Cited on page(s) xiii, 56, 68, 124, 204

[70] A. Wolf, J.B. Swift, H.L. Swinney, and J.A. Vastanoa, Determining Lyapunov exponents from a time series, *Phys. D* **16** (1985), 285–317. DOI: 10.1016/0167-2789(85)90011-9 Cited on page(s) 68

[71] The Koch snowflake, http://en.wikipedia.org/wiki/Koch_curve/. Cited on page(s) 125

[72] The Sierpinski triangle, http://en.wikipedia.org/wiki/Sierpinski_gasket/. Cited on page(s) 125

[73] Z.S. Zhang, Shift-invariant sets of endomorphisms, *Acta Math. Sinica* **27** (1984), 564–576 (in Chinese). Cited on page(s) 109

[74] Z.S. Zhang, *Principles of Differential Dynamical Systems*, Science Publishing House of China, Beijing, China, 1997. Cited on page(s) 102, 109, 124

[75] Z.L. Zhou, *Symbolic Dynamics*, Shanghai Scientific and Technological Education Publishing House, Shanghai, China, 1997 (in Chinese). Cited on page(s) xiii, 84, 109, 124

Authors' Biographies

GOONG CHEN

Goong Chen was born in Kaohsiung, Taiwan in 1950. He received his BSc (Math) from the National Tsing Hua University in Hsinchu, Taiwan in 1972 and PhD (Math) from the University of Wisconsin at Madison in 1977. He has taught at the Southern Illinois University at Carbondale (1977–78), and the Pennsylvania State University at University Park (1978–1987). Since 1987, he has been Professor of Mathematics and Aerospace Engineering, and (since 2000) a member of the Institute for Quantum Science and Engineering, at Texas A&M University in College Station, Texas. Since 2010, he is also Professor of Mathematics at Texas A&M University in Qatar at Doha, Qatar.

He has held visiting positions at INRIA in Rocquencourt, France, Centre de Recherche Mathematiques of the Université de Montréal, the Technical University of Denmark in Lyngby, Denmark, the National University of Singapore, National Taiwan University in Taipei, Taiwan, Academia Sinica in Nankang, Taiwan, and National Tsing Hua University in Hsinchu, Taiwan.

He has research interests in many areas of applied and computational mathematics: control theory for partial differential equations (PDEs), boundary element methods and numerical solutions of PDEs, engineering mechanics, chaotic dynamics, quantum computation, chemical physics and quantum mechanics. He has written over one hundred and thirty papers, five advanced texts/monographs, and co-edited four books. He is Editor-in-Chief of the Journal of Mathematical Analysis and Applications, and he has served on several other editorial boards, including the SIAM Journal on Control and Optimization, the International Journal on Quantum Information, and the Electronic Journal of Differential Equations. He is also a co-holder of a U.S. Patent on certain quantum circuit design for quantum computing.

He holds memberships in the American Mathematical Society (AMS) and the Society for Industrial and Applied Mathematics (SIAM).

YU HUANG

Yu Huang was born in Guangdong Province, People's Republic of China, in 1963. He received his BSc and MSc (Math) from Zhongshan (Dr. Sun Yat-Sen) University in Guangzhou, China, respectively, in 1983 and 1986, and his PhD (Math) from the Chinese University of Hong Kong, in Hong Kong, in 1995. He has been teaching at Mathematics Department of Sun Yat-Sen University since 1986. There, he was promoted to Professor of Mathematics in 2006.

His research interests include control theory for partial differential equations, topological dynamical systems and chaos, and switching system theory. He has written over forty papers and co-edited a book. He is an Associate Editor of the Journal of Mathematical Analysis and a guest Associate Editor of the International Journal of Bifurcation and Chaos.

Index